IN ACCORDANCE WITH THE LATEST SYLLABUS PRESCRIBED BY THE CENTRAL BOARD OF SECONDARY EDUCATION, DELHI FOR SECONDARY EDUCATION (C.B.S.E) EXAMINATION.

A COMPLETE GUIDE TO MCQ

(With complete explanations of all answer)

(C.B.S.E)

SCIENCE

(FOR CLASS X)

Author

Er. SAJAL KUMAR GHOSH
A.M.I.E

NOESIS PUBLICATION
KAIGA, KARNATAKA

Copyright©2018 Createspace, an Amazon publication unit.

For Noesis Publication, Karnataka

No part of this book may be reproduced in any form, by Photostat, microfilm, xeroxography, or any other means, or incorporated into any information retrieval system, electronic or mechanical without the written permission of the publisher.

ISBN-13: 978-1718736412
ISBN-10: 171873641X

Disclaimer: The publishers and the author or seller will not be responsible for any damage or loss of action of anyone, of any kind, in any manner therefrom. Every effort has been made to avoid errors or omissions in this publications. In spite of this, some error might have crept in. Any mistake, error or discrepancy noted may be brought to our notice which shall be taken care of in the next edition.

TO
MY WIFE JAYEETA

PREFACE

A Complete Guide to MCQ for Class X has been written in accordance with the latest syllabus of Science prescribed by the Central Board of Secondary Education (CBSE), New Delhi. The present book will help you to self-understand the subject in a better way.

Unique features of this book

- Written in very simple, easy to understand student friendly language.
- The answer is written in a comprehensive style in most cases with well-illustrated and labeled diagrams where ever it was required.
- All chapters of science book are covered.
- All questions are made at par with CBSE question standard.
- Every question has four options and they are very informative though only one will be the correct answer.
- All questions are answered at last of each chapter.
- Every answer is explained in detail.
- Other than correct answer all other options are also discussed to understand why they are not the correct option for that question.
- About 1200 Q & A.

I hope this book will prove very useful to the students and teachers.

Suggestions and constructive criticism for the further improvement of the book would be gratefully acknowledged and should incorporate in coming editions.

Er. Sajal Kumar Ghosh

Table of Contents

Chapters	Pages
Chapter 1: Chemical Reactions & Equations.	01-08
Chapter 2: Acids, Bases & Salts.	09-18
Chapter 3: Metals & Non-metals.	19-28
Chapter 4: Carbon & its Compounds.	29-38
Chapter 5: Periodic Classification of Elements.	39-49
Chapter 6: Life Processes.	50-59
Chapter 7: Control & Coordination.	60-69
Chapter 8: How do Organisms Reproduce.	70-79
Chapter 9: Heredity & Evolution.	80-90
Chapter 10: Light –Reflection & refraction.	91-100
Chapter 11: The Human Eye & the Colorful World.	101-109
Chapter 12: Electricity.	110-121
Chapter 13: Magnetic Effects of Electric Current.	122-129
Chapter 14: Source of Energy.	130-138
Chapter 15: Our Environment.	139-146
Chapter 16: Management of Natural Resources.	147-156

Chemical Reactions & Equations.

1) Chemical equations are balanced to follow

a) Law of conservation of energy.
b) Law of conservation of mass.
c) Law of conservation of momentum.
d) Both (a) and (b).

2) Substances which take part in a chemical reacion is called

a) Reactant.
b) Reagent.
c) Catalyst.
d) Product.

3) Substances which does not take part in a chemical reaction directly but have an influence in reaction they are called

a) Reactant.
b) Reagent.
c) Catalyst.
d) Product.

4) When the silver spoon is dipped in a copper sulphate solution

a) Sulfur di-oxide gas produced.
b) Copper will be precipitated.
c) Flash light will occur.
d) No reaction takes place.

5) Addition of hydrogen or metallic element is called

a) Oxidation.
b) Reduction.
c) Neutralization
d) Thermite reaction.

6) Which of the following elements never has positive oxidation state in any of its compounds

a) Fluorine.
b) Bromine.
c) Chlorine.
d) Iodine.

7) In photosynthesis ratio of a molecule of carbon di-oxide and Oxygen gas is

a) 1:2.
b) 1:1.
c) 2:1.
d) 2:3.

8) In the double displacement reaction between aqueous potassium & aqueous lead nitrate, a yellow precipitate of lead iodide is formed. In place of lead nitrate which one could be used

a) Potassium sulphate.
b) Ammonium nitrate.
c) Lead aceated.
d) Lead sulphate.

9) Silver chloride solution should
a) Stored in dark color bottle.

b) Stored in a transparent bottle.
c) Not be stored in a glass bottle.
d) Stored in a copper container.

10) Silver chloride should not be placed in sunlight because

a) The photochemical reaction takes places and it produces a silver oxide.
b) The photochemical reaction takes places and it produces silver.
c) No reaction takes place.
d) Silver and chlorine will produce due to a photochemical reaction.

11) Which of the following statement(s) is not a sign of chemical reaction?

1. Evolution of heat.
2. Evolution of light.
3. Change of color.
4. Evolution of sound.

a) 4 only.
b) 1 & 3.
c) 2 & 4.
d) None.

12) Respiration is

a) Oxidation reaction.
b) Reduction reaction.
c) Decomposition reaction.
d) Photochemical reaction.

13) NaOH + HCl = NaCl + H$_2$O Here sodium hydroxide is

a) Reduced.
b) Oxidized.
c) Decomposed.
d) Neutralized.

14) NaOH + HCl = NaCl + H2O Here hydrochloric acid is

a) Reduced.
b) Oxidised.
c) Decomposed.
d) None of the above.

15) Which of the following are combination reactions?

1. $2FeSO4 = Fe_2O_3 + SO_2 + SO_3$.
2. $4Al + 3O2 = 2Al_2O_3$.
3. $Zn + FeSO_4 = ZnSO_4 + Fe$.
4. $MgO + H_2O = Mg(OH)_2$.

a) 1 & 2 are correct.
b) 2 & 4 are correct.
c) 2, 3 & 4 are correct.
d) All are correct.

16) Addition of Oxygen or non-metallic element is called

a) Oxidation.
b) Reduction.
c) Neutralization
d) Thermite reaction.

17) Oxidizing agents are located in periodic table

a) Mostly iat the right side
b) Mostly in the middle.
c) Mostly at left side.
d) No specific location observed.

18) Antioxidants are used in food for

a) Prevent food from adulteration.
b) Improving the taste of food.
c) Protecting food from rancidity.
d) All of the above purpose.

19) Most of the decomposition reactions are

a) Endothermic reaction.
b) Exothermic reaction.
c) Oxidation reaction.
d) Reduction reaction.

20) Most of the combination reactions are

a) Endothermic reaction.
b) Exothermic reaction.
c) Oxidation reaction.
d) Reduction reaction

[A COMPLETE GUIDE TO MCQ]   CHAPTER: 01

21) NaOH + HCl = NaCl + H2O , it is an example of

a) Displacement reaction.
b) Double displacement reaction.
c) RedOx reaction.
d) Both (b) and (c).

22) Respiration is

a) Neutralization reaction.
b) Reduction reaction.
c) Combination reaction.
d) Photochemical reaction.

23) Loss of electron is called

a) Reduction.
b) Oxidation.
c) Neutralization.
d) Dilution.

24) 2KClO3 +Δ + Catalyst = 2KCl +3O2.

a) It is a combination and endothermic reaction.
b) It is a catalytic exothermic reaction.
c) It is an endothermic decomposition reaction.
d) It is a photochemical reduction reaction.

25) Where Which of the following statement is correct for following reaction:

$3Fe + 4H_2O = Fe_3O_4 + 4H_2$

a) Iron being oxidized.
b) Water being reduced.
c) Iron is acting as reducing agent.
d) Water is acting as oxidizing agent.

26) Which of the following is(are) endothermic process(es)

1. Sublimation of dry ice.
2. Condensation of water vapour.
3. Evaporation of water.
4. Dilution of sulphuric acid in water.
5. Solidification of water.

a) 1 & 3 are correct.
b) 2 & 3 are correct.
c) 2 & 4 are correct.
d) 1 & 5 are correct.

27) The thermite reaction is an example of

a) Exothermic and decomposition reaction.
b) Endothermic and composition reaction.
c) Endothermic and displacement reaction.
d) Exothermic and double displacement reaction.

28) The following reaction is an example of a
   4NH3+5O2 = 4NO + 6H2O

a) Combination reaction.
b) Red ox. reaction.
c) Neutralization reaction.
d) Displacement reaction.

29) For long term edible oil storage which of the following gas is used

1. Hydrogen.
2. Helium.
3. Nitrogen.
4. Carbon di-oxide.

a) 1 & 2 are correct.
b) 2 & 3 are correct.
c) 2 & 4 are correct.
d) All are correct.

30) The most effective reducing agent is

a) Li
b) Mg
c) H
d) Na

31) Antioxidants used in the following type of food for long time preservation.

a) The food contains Fats & Protein only
b) The food contains Protein only.

Chemical Reactions & Equations MCQ (class X) [Page 3]

c) The food contains carbohydrates only
d) The food mainly contains Fats & oil

32) When water gets electrolyze the mole ratio of Hydrogen & Oxygen gases is

a) 1:1.
b) 1:2.
c) 2:1.
d) 2:3.

33) Number of water molecule present in one molecule of Copper sulphate is

a) 3
b) ½
c) 5
d) 7

34) $2Pb(NO_3)_2 = 2PbO + O_2 + X$

What is X in this given reaction?

a) 4NO
b) $4NO_2$
c) NO_2
d) $2PbNO_2$

35) The reducing agent used to obtain manganese from manganese dioxide is

a) Fluorine.
b) Iodine.
c) Carbon
d) Hydrogen.

36) The oxidation reaction which produces heat & light is called

a) Combustion.
b) Exothermic.
c) Endothermic.
d) Photochemical.

37) $MnO_2 + 4HCl = MnCl_2 + A + Cl_2$ in this reaction A is

a) H_2O
b) $2 H_2O$
c) $3 H_2O$
d) $4 H_2O$

38) Quick lime and slaked lime are following compounds.

a) $Mg(OH)2$ and $Ca(OH)2$.
b) CaO & $Ca(OH)2$.
c) $Ca(OH)2$ & CaO.
d) $Ca(OH)2$ & $Mg(OH)2$.

39) Oxidizing agents are those which

a) Supply oxygen.
b) Removes oxygen.
c) Supply hydrogen.
d) remove electrons.

40) Dilute Sulfuric Acid solution produced by

a) Adding Water in Conc. H2SO4.
b) Adding Conc. H_2SO4 in water.
c) Adding equal quantity water and Conc. H2SO4 in a bottle.
d) All of the above methods are applicable.

41) Bio-gas main component on burning produces

a) SO_2 and CO_2
b) CO_2 and Water.
c) CO_2
d) H_2O

42) $Cu + aHNO3 \rightarrow Cu(NO3)2 + bNO2 + 2H2O$ here value of a & b is

a) 4 and 3 respectively.
b) 1 and 3 respectively.
c) 3 and 4 respectively.
d) The reaction is not possible.

43) Reducing agents are those which

a) Supply oxygen.
b) Removes oxygen.
c) Supply hydrogen.
d) Supply electrons.

44) When magnesium ribbon burns in the air it produces following compounds

a) Mg3N2 & Mg(OH)2
b) MgO.
c) Mg_3N_2 & MgO.
d) MgCO3 & MgO.

45) Heating of green vitriol gives which of the following product?

a) Ferric oxide.
b) Sulfur dioxide.
c) Sulfur trioxide.
d) All of the above.

46) The burning of green vitriol smelled as

a) Burning magnesium ribbon.
b) Burning sulpher.
c) Burning of cellulose.
d) Burning of carbon.

47) Magnesium ribbon is cleaned before burning to

a) Remove magnesium nitride.
b) Remove magnesium oxide.
c) Remove magnesium carbonate.
d) Remove magnesium chloride.

48) NaOH +HCl =NaCl +H2O here NaOH is

a) Reduced.
b) Oxidized.
c) Reducing agent.
d) In gaseous form.

49) NaOH +HCl =NaCl +H2O here HCl is

a) Reduced.
b) Oxidized.
c) Oxidizing agent.
d) In gaseous form.

50) Conditions of a chemical reaction are written

a) On above / below the arrow mark of the equation.
b) At the left side of the equation.
c) At the right side of the equation.
d) Not required to mention in the equation.

51) When Magnesium ribbon burns in the air it

a) Oxidized.
b) Reduced.
c) Neutralized.
d) Both (a) and (b)

52) The chemical name of slaked lime is

a) Calcium oxide.
b) Calcium hydroxide.
c) Calcium carbonate.
d) Sodium hydroxide.

53) Burning of fuel in the engine is an example of

a) Oxidation.
b) Reduction.
c) Both (a) and (b)
d) Displacement reaction.

54) Which statement(s) are not correct regarding a chemical reaction?

1. All reactions equation required balancing.
2. By equating a number of the molecule, we equate mass.
3. All chemical reaction takes place in a certain condition.
4. All reaction produces energy.

a) 1 & 4 are wrong.
b) 2 & 4 are wrong.
c) 1 & 3 are wrong.
d) 2 & 3 are wrong.

55) The color of magnesium oxide is

a) Green.
b) Black.
c) Yellow.
d) White.

56) During balancing of a chemical equation, we should first try to balance those molecules whose number of atoms

a) Minimum.
b) Average.
c) Maximum.
d) Any one element could take for consideration.

57) Burning of coal in the air is an example of

a) Combination reaction.
b) Decomposition reaction.
c) Displacement reaction.
d) Double displacement reaction.

58) Burning of methane in air is an example of

a) Combination reaction.
b) Decomposition reaction.
c) Exothermic reaction.
d) Endothermic reaction.

59) Photosynthesis is an example of

a) Combination reaction.
b) Decomposition reaction.
c) Exothermic reaction.
d) Endothermic reaction.

60) When lead nitrate is heated a brown color fume is produced. This fume is

a) NO_2
b) NO
c) N_2O
d) N_2

61) Barium sulphate is

a) Soluble in water.
b) Insoluble in water.
c) Slightly soluble in water.
d) Produce colloidal solution.

62) When Hydrogen gas is passed over the heated copper oxide _____ color metal remains due to _____.

a) Black, Oxidation.
b) Red, Oxidation.
c) Red, Oxidation.
d) Red, Reduction.

[A COMPLETE GUIDE TO MCQ]

CHAPTER: 01

[Answers & Explanations]

1) **(d).** Mass and energy both are conserved in this universe.

2) **(a).** reactant are substances that take part in and undergoes change during a chemical reaction.

3) **(c). Catalyst** is those substances which do not take part in a chemical reaction but it changes the reaction rate (either increase or decrease). When it increases reaction rate it is called POSITIVE CATALYST and if it reduces reaction rate then it is called NEGATIVE CATALYST.

4) **(d).** Silver placed below the copper in reactivity series so no reaction will take place.

5) **(b).** Addition of Hydrogen or electropositive element called reduction.

6) **(a).** Fluorine has the highest affinity to the electron.

7) **(b).** During photosynthesis for production of each molecule of glucose 6 molecule CO_2 & 6 molecule water is used which produce 6-molecule oxygen as by product.
   $$6CO_2 + 6H_2O = C_6H_{12}O_6 + 6O_2$$

8) **(c).**

9) **(a).** Silver chloride decomposes to Silver and Chlorine due to the presence of sunlight so it is stored in dark color bottle.

10) **(d).** See Q-9

11) **(d).** all are a sign of chemical reaction.

12) **(a).** During respiration food oxidized in presence of oxygen supplied by the blood.

13) **(a).** NaOH due to gain of an electron.

14) **(b).** HCl oxidized due to loss of an electron.

15) **(b).** Sl.no: 1 Decomposition reaction. Sl.no: 2 & 4: Combination reaction. Sl.no: 3 Displacement reaction.

16) **(a).** As per definition.

17) **(c).** O, F, Cl, S all are oxidizing agents.

18) **(c).** Antioxidants protect food from oxidation i.e. rancidity.

19) **(a).** All decomposition reaction required heat so they are an endothermic reaction.

20) **(b).** Combination reaction due to the formation of new bonds they release heat so they are an exothermic reaction.

21) **(d).** As per definition both (b) and (c) applicable.

22) **(c).** During respiration, Oxygen combines with the food.

23) **(b).** Loss of electron is called oxidation reaction.

24) **(c).** As heat is added at the right side of the equation so it is called endothermic reaction and due to decomposition to KCl and O_2 it is also a decomposition reaction.

25) **(a).** Iron combined with oxygen so it is oxidized.

26) **(a).** Sublimation & Evaporation require heat from outside so it is an endothermic reaction. All other process releases heat.

27) **(c).** Thermite reaction require heat to add so this is a Endothermic reaction and as O_2 is displaced from iron and oxidized aluminum so it's a displacement reaction.
   $$Fe_2O_3 + 2\ Al \rightarrow 2\ Fe + Al_2O_3$$

[A COMPLETE GUIDE TO MCQ]

CHAPTER: 01

28) **(d).**

29) **(b).** Helium and Nitrogen protect food from oxidation. These gases to protect food from rancidity flush food items.

30) **(c).** Hydrogen located at first period and first group.

31) **(d).** Food contain fat and oils are prone to rancidity.

32) **(c).** $2H_2O = 2H_2 + 1O_2$

33) **(c).** $CuSO_4, 5H_2O$.

34) **(b).** Try to balance the equation.

35) **(c).** Calcination method.

36) **(a).** Combustion produces both heat and light.

37) **(b).** Balance the chemical equation.

   $MnO_2 + 4HCl = MnCl_2 + 2H_2O + Cl_2$

38) **(b).** CaO is quick lime and $Ca(OH)_2$ is slaked lime.

39) **(a).** As per definition.

40) **(b).** Dilute acid solution produced by addition of conc. Acid with water.

41) **(b).** Bio gas main component is Methane (CH_4) and on burning is produces CO_2 and H_2O.

   $CH_4 + 2O_2 = CO_2 + 2H_2O$

42) **(c).** Balance the equation.

43) **(b).** As per definition of oxidation.

44) **(c).** When magnesium burn in atmosphere it react with nitrogen and oxygen both so it produces Mg_3N_2 and MgO both.

45) **(d).** $2FeSO_4 = Fe_2O_3 + SO_2 + SO_3$

46) **(b).** Due to production SO_2 gas.

47) **(b).**

48) **(a).**

49) **(b).**

50) **(a).**

51) **(a).** Magnesium oxidized and form MgO

52) **(b).** $Ca(OH)_2$

53) **(a).** During burning, fuel constituents oxidized.

54) **(a).** Not all chemical equation require balancing as they itself balanced such as
   $Zn + H2SO4 = ZnSO4 + H2$
   and endothermic reactions are absorb energy, they does not produces energy e.g. Photosynthesis.

55) **(d).**

56) **(c).**

57) **(a).** During burning, substances are oxidized.

58) **(c).** Burning of methane gas produces heat energy.

59) **(d).** Photosynthesis stored energy so its endothermic reaction.

60) **(d).** $2P(NO_3)_2 = 2PbO + 4NO_2 + O_2$

61) **(b).**

62) **(d).** Copper metal color is red and it's happened due to reduction of CuO to Cu.

Acids, Bases & Salts

1) Which acid is used in cola to give it a biting sharp taste

a) Sulphuric acid (H_2SO_4)
b) Tartaric acid.
c) Phosphoric acid.
d) Citric acid.

2) Range of pH scale is

a) 7 to 10
b) 0 to 10
c) 0 to 14
d) 7 to 14

3) Due to excess passing of CO2 through aqueous solution of slaked lime, its milkiness fades because

a) Calcium carbonate is produced.
b) Calcium bi-carbonate is produced
c) Calcium oxide is produced.
d) Due to production of more heat.

4) When acids dissolves in water it releases

a) H^+ ion.
b) H^- ion.
c) H_3O^+ ion.
d) $H_3O_2^+$ ion.

5) Which element is always present in Arrhenius acid

a) Oxygen.
b) Nitrogen.
c) Hydrogen.
d) None of the above.

6) Methyl red color in acidic medium is

a) Yellow.
b) Pink.
c) Red.
d) Orange.

7) Chemical formula of Gypsum is

a) $CaSO_4, 1/2H_2O$
b) $CaSO_4, 2H_2O$
c) $CaSO_4, H_2O$
d) $CaSO_4, 3H_2O$

8) During preparation of HCl gas on a humid day, the gas is usually pass through the guard tube containing CaCl2. The purpose of using CaCl2 is

a) To add moisture to the gas (HCl).
b) To absorbe HCl gas.
c) To absorb moisture from HCl gas.
d) Used as a catalyst.

9) Which one is different from others

a) Nitric acid.
b) Sulphuric acid.
c) Tartaric acid.
d) Phosphoric acid.

Acids, Bases & Salts MCQ (class X)

10) Soda ash chemical formula is

a) NaHCO3.
b) $Na_2CO_3, 9H_2O$.
c) Na_2CO_3.
d) $Na_2CO_3, 10H_2O$.

11) Common salt beside being used in kitchen can also be used as the raw material for production of

1. Backing powder.
2. Washing soda.
3. Black ash.
4. Slaked lime.

a) 2 & 3
b) 1 & 3
c) 1 & 2
d) 2 & 4

12) Antacid solutions pH found

a) ≤6.5
b) ≥7.0
c) >10
d) >14

13) Black ash is

a) Dry KOH.
b) Dry washing soda.
c) Charcoal.
d) Hydrated KOH.

14) When electricity passes through NaCl aqueous solution

a) Sodium metal is deposited.
b) Only Chlorine gas produced.
c) Chlorine & Hydrogen gases are produced.
d) The entire above are produced.

15) Identify the element

1. It is white translucent solid.
2. Its readily reacts with water and produces alkaline solution.
3. It is stored in kerosene.

a) Na.
b) Ca.
c) Al.
d) P.

16) Bleaching powder chemical name is

a) Calcium hypo-Oxichloride.
b) Calcium oxy-Chloride.
c) Calcium Chloride.
d) Calcium Chloro-Oxide.

17) Methyl orange color in basic medium is

a) Pink.
b) Orange.
c) Purple.
d) Yellow.

18) The pH of toothpaste commonly used is

a) <6.5
b) ≥7.0
c) ≥2.2
d) None.

19) Phenolphthalein color in basic medium is ____ but in acid it is ____.

a) Pink, Colorless.
b) Yellow, Pink.
c) Pink, Orange.
d) Blue, Red.

20) When sodium hydroxide reacts with Zinc it produces

a) Sodium oxide and water.
b) Sodium zincates and water.
c) Sodium zincates and hydrogen.
d) Sodium oxide and hydrogen.

21) You are given 3 unknown solutions with pH value as 6, 8 & 9.5 respectively. Which solution will contain maximum OH- ion

a) Solution sample-1.

b) Solution sample-2.
c) Solution sample-3.
d) Data insufficient.

22) The pH of Gastric juice is

a) <6.5
b) ≥7.0
c) ≤2.2
d) None.

23) Chemical formula of Plaster of Paris is

a) $CaSO_4, 1/2H_2O$
b) $CaSO_4, 2H_2O$
c) $CaSO_4, H_2O$
d) $CaSO_4, 3H_2O$

24) Ratio of water molecule in Plaster of paris & Gypsum is

a) 3:1
b) 1:3
c) 1:4
d) 4:3

25) Ammonium sulphate salt is

a) Basic salt.
b) Acidic salt.
c) Neutral salt.
d) Complex salt.

26) Which of the following acid(s) never form acidic salt?

1. HCl
2. H_3PO_4
3. H_2SO_4
4. H_2CO_3

a) 1 only.
b) 4 only.
c) 1 & 4 both.
d) 2 & 3 both.

27) Vinegar is used in pickling as it

a) Is an acid.
b) Prevents growth of microbes.
c) Prevent drying of pickle.
d) Increase taste.

28) pH scale of neutral solution is

a) 14.
b) 7.
c) 10
d) 12.

29) Which of the following acids are edible

1. Citric acid.
2. Tartaric acid.
3. Hydrochloric acid.
4. Carbonic acid.

a) 1 & 2 are correct.
b) 1,2 & 4 are correct.
c) 1,2 & 3 are correct.
d) All are correct.

30) Butyric acid is found in

a) Rancid butter.
b) Rancid cake.
c) Stings of bees.
d) All of the above.

31) An indicator is one kind of following compound

a) Strong acid only.
b) Reducing agent.
c) Weak base or acid only.
d) Complex salt.

32) NaCl is a

a) Basic salt.
b) Acidic salt.
c) Neutral salt.
d) Complex salt.

33) Which of the following phenomena occurs when acid mixed with water

1. Neutralization.
2. Dilution.
3. Ionization.

a) Only 2 is correct.
b) 1 & 2 are correct.
c) 2 & 3 are correct.
d) Only 3 is correct.

34) Baking powder is

a) Sodium carbonate + Sodium tartaric.
b) Sodium bi-carbonate + Sodium tartaric.
c) Sodium carbonate + Tartaric acid.
d) Sodium carbonate + Sodium benzoate.

35) Which of the following contains Oxalic acid?

a) Tomato
b) Lemon
c) Grape
d) Orange.

36) Milk of magnesia is

a) $Mg_2(OH)_3$
b) MgO
c) $Mg(OH)_2$
d) Mg_3N_2.

37) In chemical laboratory most common indicator is

a) Methyl Orange.
b) Phenolphthalein.
c) Methyl red.
d) Universal indicator.

38) Salt conduct electricity when

a) It is in solution state in water.
b) It is in dry crystallize state.
c) It is in molten state.
d) Both (a) and (c).

39) When acids or bases are diluted with water it is

a) Exothermic reaction.
b) Endothermic reaction.
c) Combination reaction.
d) Neutralization reaction.

40) Gastric juice contains HCL which is one example of

a) Inorganic acid.
b) Organic acid.
c) Soft organic acid.
d) Strong inorganic acid.

41) When acids reacts with metal oxide it produces

a) Water and salt.
b) Salts and hydrogen gas.
c) Salts only.
d) No reaction takes place.

42) Which of the following used to remove permanent hardness of water?

a) Sodium carbonate.
b) Sodium bi-carbonate.
c) Calcium chloride.
d) Magnesium sulphate.

43) Green vitriol is (crystal form)?

a) Ferrous sulphate.
b) Copper sulphate.
c) Aluminum sulphate.
d) Calcium chloride.

44) Which of the following is the most strong acid?

a) Nitric acid.
b) Sulphuric acid.
c) Phosphoric acid.
d) Hydrochloric acid.

45) When more and more water in diluted with acids its H+ ion concentration will

a) Increases.
b) Decreases.
c) Remain same.
d) Depends on type of acids.

46) Which alkaline used as antacid?

a) $Al_2(OH)_3$

b) MgO
c) Mg(OH)$_2$
d) KOH.

47) Which of the following could use as an olfactory indicator?

a) Clove.
b) Vanilla.
c) Onion.
d) All of the above.

48) Blue vitriol is.

a) Ferrous sulphate.
b) Copper sulphate.
c) Aluminum sulphate.
d) Calcium chloride.

49) Yellow color turmeric turns into _____ in acid solution

a) Yellow.
b) Red.
c) Purple.
d) Blue.

50) Yellow color turmeric turns into _____ in alkaline solution

a) Yellow.
b) Red.
c) Purple.
d) Blue.

51) Olfactory indicators are those

a) Color changes in acid and base medium.
b) Odour changes in acid and base medium.
c) Color changes in acid but not in base medium.
d) Color changes in base but not in acidic medium.

52) When milk of magnesia reacts with acetic acid it produces

a) Basic salts
b) Acidic salt.
c) Neutral salt.
d) Complex salt

53) Human body function correctly when pH range of blood is

a) 6.8-7.8
b) 7-7.8
c) 6-8
d) 5.5-8.5

54) Phosphoric acid is

a) Monoprotic acid
b) Diprotic acid
c) Triprotic acid
d) polyprotic acid

55) pH of soda water is

a) >5
b) <10
c) >7
d) <7

56) Alum is a

a) Mixed salt.
b) Acidic salt.
c) Basic salt.
d) None of the above.

57) Which of the following phenomena will occur when a small amount of acid is added to water?

a) Dilution.
b) Neutralization.
c) Salt formation.
d) Ionization.

58) Only sodium bi-carbonate is not used as baking powder because

a) It produces sodium carbonate.
b) Production of CO2 by heating alone sodium bi-carbonate is requires high-energy supply.
c) It makes poisonous food.
d) It works very slowly.

59) Chlor-Alkali process is used to produce

a) KOH
b) NaOH
c) Aluminum.
d) Sodium.

60) Brine is used for industrial production of

a) NaOH.
b) KOH.
c) Bleaching powder.
d) None of the above.

61) Which gas is used for producing "margarine"

a) Nitrogen.
b) Carbon di-oxide.
c) Hydrogen.
d) Chlorine.

62) In Chlor-Alkali process the byproduct gas(s) are?

a) Hydrogen only.
b) Hydrogen and Oxygen gas.
c) Hydrogen and chlorine gas.
d) Chlorine and Nitrogen gas.

63) Table salt is a raw material for production of following

1. NaOH.
2. Washing soda.
3. Baking powder.

a) 1 & 2.
b) 1 & 3.
c) 1, 2 & 3.
d) 2 & 3.

64) Which one of the following is used for manufacturing bleaching powder?

a) Hydrogen.
b) Oxygen.
c) Nitrogen.
d) Chlorine.

65) Dehydration reaction is one type of

a) Exothermic reaction.
b) Endothermic reaction.
c) Photochemical reaction.
d) Combination reaction.

66) On heating, acidic carbonate salt of sodium will give

a) CO.
b) CO_2.
c) SO_2
d) H_2

67) Which of the following is used as weed killer?

a) Black ash.
b) White ash.
c) Blue vitriol.
d) Calcium chlorohypochloride.

68) To degrease a metal you should use

a) Sodium hydroxide.
b) Calcium chloride.
c) Hydrogen gas.
d) Sulpher di-oxide.

69) Baking powder used

a) Basic salt of sodium.
b) Acidic salt of sodium.
c) Acidic salt of calcium.
d) Neutral salt of sodium.

70) Following salt is not hygroscopic in nature

a) NaCl
b) MgCl
c) $CaCl_2$
d) KCl

71) When a base reacts with non-metal oxide

a) It's neutralized each other.
b) Its creates fire.
c) Its produces acidic salts.
d) Its produces basis salts.

72) Acidic salt produced when

a) Weak acid reacts with strong alkali.
b) Weak alkali reacts with strong acid.
c) Weak acid reacts with weak alkali.
d) Strong alkali reacts with strong acids.

73) Basic salt produced when

a) Weak acid reacts with strong alkali.
b) Weak alkali reacts with strong acid.
c) Weak acid reacts with weak alkali.
d) Strong alkali reacts with strong acids.

74) One name of common salt is

a) Oil of Vitriol.
b) White vitriol.
c) Table salt.
d) Green vitriol.

75) Burning of fossil fuel releases following oxides

1. Oxide of sulpher.
2. Oxide of nitrogen.
3. Oxide of phosphorous.
4. Oxide of chlorine.

a) 1 & 2 only.
b) 1 & 3 only.
c) 2 & 3 only.
d) 1,2 and 3.

76) Ammonium shulphate is a.

a) Mono basic salt.
b) Mono acidic salt.
c) Neutral salt.
d) Tetra acidic salt.

77) pH value of neutral salt is

a) 7
b) <7
c) >7
d) 0

78) Brine is

a) Solution of calcium chloride.
b) Solution of zinc chloride.
c) Solution of sodium chloride.
d) Solution of magnesium chloride.

79) Acid rain is harmful for Tajmahal because it is made of

a) Calcium sulphate.
b) Calcium carbonate.
c) Calcium oxide.
d) Calcium chloride.

80) When Sulphuric acid react with egg shell it produces _____ gas

a) Hydrogen.
b) Nitrogen.
c) Carbon monoxide.
d) Carbon di-oxide.

[A COMPLETE GUIDE TO MCQ]

CHAPTER: 02

[Answers & Explanations]

1) **(c)**. Phosphoric acid one of the inorganic edible acid used in cola to give biting sharp taste.

2) **(c)**.

3) **(b)**. Due to excess passing of CO2 gas slaked lime i.e. insoluble calcium carbonate converted to soluble calcium bi-carbonate so its milkiness fades.

4) **(a)**. According to **Arrhenius theory** acids releases H+ ion.

5) **(c)**. Hydrogen is the element, which always present as according to **Arrhenius theory** acids releases H+ ion.

6) **(a)**. Yellow. In basic medium its color is Red.

7) **(b)**. Gypsum is produced by hydrating plaster of paris.

   $CaSO_4, \frac{1}{2}H_2O + 1\frac{1}{2} H_2O = CaSO_4, 2H_2O$

8) **(c)**. $CaCl_2$ is hygroscopic salt so it will absorb moisture produced during the reaction so that dry HCL could be produced. otherwise, it will become Hydrochloric acid.

9) **(c)**. **Tartaric acid** is organic acid and all other are inorganic acid.

10) **(c)**. Soda ash is anhydrate sodium carbonate.

11) **(c)**. Black ash (dry KOH) and slaked lime ($CaCO_3$) does not contain sodium.

12) **(b)**. Antacids is bases.

13) **(a)**. Dry KOH. Dry Washing soda is soda ash. [Note Q-10.]

14) **(c)**. Its **Chlor-Alkali** process of producing NaOH from NaCl.

15) **(a)**. Sodium stored in kerosene and readily react with water to form alkaline solution. None other is such stored in kerosene and readily react with water.

16) **(b)**. It's also known as Chloro hypo chloride.

17) **(b)**. Methyl orange and methyl red both name indicates color in basic medium.

18) **(b)**. Tooth decay occurs due to acids. Toothpaste neutralizes the acids produce by bacteria with reaction of remaining food inside the mouth.

19) **(a)**.

20) **(c)**. **2NaOH+ Zn = Na$_2$ZnO$_2$ + H$_2$**.

21) **(c)**. Sample-1 is acid (pH<7) so there is no question of OH-. Sample-2 and 3 are basic as pH is more than 7. pH of third sample is maximum and its OH- concentration will be maximum.

22) **(c)**. Its highly acidic due to presence of HCl.

23) **(a)**. See Q-7.

24) **(c)**. Plaster paris having water molecule $=\frac{1}{2}$ & Gypsum contains 2 water molecule. So ratio = 0.5/2=0.25=1:4

25) **(c)**. Salts which produces H+ or OH- ion when dissolve in water is called acidic or basic salt respectively. Here Ammonium sulphate salt (NH4Cl) does not contain any such hydrogen atom, which could produce H$^+$ ion or OH$^-$ ion and its neutral salt. [NH$_4$Cl = NH$_4^+$ + Cl$^-$]

26) **(c)**. **Acid salts** can be formed only with neutralization of Diprotic or polyprotic **acids**. **Hydrochloric acid** has only one hydrogen atom that is displaced during the reaction and therefore **hydrochloric acid salts** are **not acidic** in nature. Carbonic

Acids, Bases & Salts MCQ (class X)

acid completely displaced its hydrogen atom during reaction so it never forms acidic salts.

27) **(b).** Being a mild acid it prevents the growth of microbes.

28) **(b).** For a neutral solution, pH is 7. If its more than 7 then its basic and if its less than 7 then its acidic. Total pH scale range is 0 to 14.

29) **(c).** Citric and tartaric acid are from organic body lemon and tamarind respectively and they are edible. Hydrochloric acid though formed inside stomach it is not edible from outside the body. Carbonic acid is mild acid and is edible in the form of soda water.

30) **(a).** Rancid butter produces Butyric acid.

31) **(c).** Weak base and acids used as an indicator.

32) **(c).** NaCl not contains any H atom and not produces any H+ ion when dissolving in water. It's a neutral salt.

33) **(c).** Neutralization occurs when its reacts with alkali or base.

34) **(b).** Baking powder is a combination of sodium bi-carbonate and tartaric acid. Though only sodium bi-carbonate could be used due to the bitter test of sodium carbonate which is formed due to decomposition, its not used independently. Sodium tartrate has a pleased taste.

35) **(a).** lemon –Citric acid. Grape-Acetic acid, Orange –Citric acid. Tomato- Oxalic acid.

36) **(c).** Mg(OH)2 is called milk of magnesia and used as an antacid compound.

37) **(d).** It's a mixture of dyes that change color gradually over a range of pH and is used (especially as indicator paper) in testing for acids and alkalis

38) **(d).** Dry crystalize state does not produce free ions so it does not conduct electricity.

39) **(a).**

40) **(a).** Though it produces inside stomach, it is not organic acid as its all components are from minerals and its mineral/ inorganic acid. Its strong acid also.

41) **(a).** Metallic oxides are alkaline in nature and it produces water & salt when reacts with acid.

42) **(a).**

43) **(a). Crystalline Ferrous sulphate** is called green vitriol due to its green color.

44) **(a).** Nitric acid is the strongest acid along the examples.

45) **(b).** due to more dilution, an acid solution becomes the weak solution.

46) **(c)** Mg(OH)$_2$ is called milk of magnesia and used for antacid as it's a soft alkaline substance.

47) **(d).** All examples are changing their smell in acidic and basic solutions.

48) **(b).** Crystalline Copper Sulphate is known as blue vitriol due to its blue color.

49) **(a).** Turmeric does not change color in acidic medium but changes to red color in basic medium.

50) **(b).** See Q-49.

51) **(b).**

52) **(c)** Milk of magnesia is Magnesium Hydroxide [Mg(OH)2] and it's a softalkali. Acetic acid is soft acid. Soft acid and soft alkaline produce neutral salt.

Acids, Bases & Salts MCQ (class X)

53) (b). Blood pH range is 7-7.8.

54) (c). H_3PO_4 contains 3 hydrogen atom which could produce 3 H^+ ion in solution.

55) (d). Soda water contains Carbonic acid [H2CO3] and its pH should < 7 in pH scale.

56) (a). Alum is **$K_2SO_4, Al_2(SO_4)_3, 24H_2O$**

57) (a) Dilution only, as the production of H^+ ion decrease, when a small amount of water added in water. Neutralization possible when alkali used.

58) (a). Sodium carbonate spoil taste of food.

59) (b) Radioactivity is not a periodic property.

60) (a) Brine is a solution of sodium chloride [NaCl]. Its use in the Chloro-Alkali process to produce NaOH.

61) (c) The hydrogenation process is used.

62) (c) 2NaCl + 2H2O —> 2NaOH + Cl2 + H2

63) (a) The chlor-alkali process used for the production of NaOH
[2NaCl+2H$_2$O—>2NaOH+**Cl$_2$**+ **H$_2$**]
and Solvay process used for Washing soda production.
2NaCl+CaCO$_3$= CaCl$_2$+**Na$_2$CO$_3$**.

64) (d) **$Ca(OH)_2 + Cl_2 = CaOCl_2 + H_2O$**

65) (b) For dehydration heat required from outside for heating.

66) (b) Sodium Bi-carbonate is an acidic salt and on heating, its give CO$_2$ gas.
2NaHCO$_3$ =Na$_2$CO$_3$ + CO$_2$ + H$_2$O

67) (d) Calcium Chlorohypochloride is bleaching powder and is used for weed killing.

68) (a)

69) (b) Sodium Bi-carbonate is an acidic salt and its use in baking powder.

70) (a) A hygroscopic substance is one that readily attracts water from its surroundings, through either absorption or adsorption. Except for NaCL all are hygroscopic in nature. Si, Ge, As, Te and Po are the metalloids in the periodic table.

71) (a) Non-metallic oxide is acidic in nature and when reacts with base its neutralize each other.

72) (b).

Weak acid + Weak base =Neutral salt.
Strong acid +Strong base =Neutral salt.
Weak acid +Strong base =Basic salt.
Strong acid + Weak base =Acidic salt.

73) (a). See Q-73.

74) (c).

75) (a). Fossil fuels mainly contain Sulpher and Nitrogen. On burning, they oxidize.

76) (c).

77) (a). Neutral solution **pH** is **7.**

78) (c). NaCl solution is called **Brine.**

79) (b). Marble is Calcium Carbonate.

80) (b). Eggshell contain calcium carbonate and when reacts with H$_2$SO$_4$ it produces CO$_2$ gas.

Metals & Non-metals

1) The ability of the wire to drawn into a wire is called

   a) Malleability.
   b) Hardness.
   c) Ductility.
   d) Sonorousity.

2) **Aqua regia related to**

   a) A mixture of two soft acids.
   b) A mixture of two soft bases.
   c) A mixture of acidic and basic salt solution.
   d) A mixture of two strong acids.

3) **Which of the following alloy having one of the components as non-metal?**

   a) Silver amalgam.
   b) Brass.
   c) Bronze.
   d) Stainless steel.

4) **Galvanization is a method to**

   a) Protect iron metal from corrosion.
   b) Extract iron forms its ore.
   c) Protect food from rancidity.
   d) Improve the ductility property of metal.

5) **Which of the following acid does not give hydrogen gas on reacting with metals?**

   a) HNO_3.
   b) HCl
   c) H_2SO_4

   d) CH_3COOH

6) **Which of the following metal is liquid at body temperature?**

   a) Os (osmium)
   b) Br (Bromine).
   c) Ga (Gallium)
   d) Hg (mercury).

7) **The constituents of Aqua Regia is**

   a) Conc. H_2SO_4 and Conc. HCl
   b) Conc. H_2SO_4 and Conc. HNO_3
   c) Conc. HNO_3 and Conc. HCl
   d) Conc. H_2SO_4 and aquas HNO_3

8) **The ratio of acids (H_2SO_4: HNO_3) in Aqua regia is**

   a) 1:3
   b) 2:3
   c) 3:1
   d) 2:1

9) **Which of the following nonmetal is lustrous?**

   a) S
   b) Br
   c) I
   d) C

10) **Following metals are catch fire in the air**

   1. Al

2. Na
3. K
4. Ca

a) 2 & 3
b) 1 & 3
c) 1 & 2
d) 2 & 4

11) The ability of metals to be drawn into thin wire is known as

a) Malleability.
b) Sonority.
c) Conductivity.
d) Ductility

12) Silver articles become black on prolong exposure due to the formation of

a) Silver nitride.
b) Silver sulfide.
c) Silver oxide.
d) Silver chloride.

13) Iron is extracted from the following ore

a) Cinnabar.
b) Hematite.
c) Bauxite.
d) Calamine.

14) During electrolysis refining of a copper pure copper rod connected with

a) Positive pole.
b) Negative pole.
c) Any type of pole could be used but the negative pole is preferable.
d) Copper not refined by electrolysis.

15) Zinc blende is

a) Sulphate ore from Zinc.
b) Sulphide ore of Zinc.
c) Carbonate ore of Zinc.
d) Oxide ore of Zinc.

16) In thermite process, Al metal used as

a) Oxidizing agent.
b) Reducing agent.
c) Positive catalyst.
d) Negative catalyst.

17) The following acid could dissolve gold

a) Conc. HNO3.
b) Conc. H3PO4
c) Aqua regia.
d) Milk of magnesia

18) Stainless steel required following metals as its component

a) Mn, Mg and Cr
b) Mn, Mg and Cr
c) Mn, Ni and Cr
d) Mn, Cu and Cr

19) When Al powder is heated with MnO2, MnO2 getting

a) Oxidized.
b) Reduced.
c) Neutralized.
d) No reaction takes place.

20) For joining rail lines following method is used

a) Thermite.
b) Riveting.
c) Welding.
d) Anodizing.

21) During calcinations process, Oxygen supply is maintained

a) Excess.
b) Limited.
c) As same as reaction required.
d) Oxygen never used in this process.

22) If Al is dipped in Conc. HNO3 then its reactivity

a) Increases.

b) Decreases.
c) Remain constant.
d) No reaction takes place.

23) Malleability is the property of metal due to which it could be easily

a) Form wire.
b) Form thin sheet.
c) Form acidic oxide.
d) Easily catch fire in the air.

24) Which of the following metal alloy contain Hg as an element

a) Alnico.
b) Hindalco.
c) Solder.
d) Silver amalgam

25) Zinc blende is

a) Zinc sulphide.
b) Zinc Chloride.
c) Zinc Carbonate.
d) Zinc sulphate

26) Which of the following is/are not an ionic compound?

1) HCl gas   2) CCl4   3) KCL   4) MgCl2

a) 1 & 2.
b) 2 only.
c) 2 & 4.
d) 1 & 3.

27) Following non-metal is a good conductor of electricity and heat both

a) Graphite.
b) Fullerene.
c) Diamond.
d) Sulpher.

28) Reactivity of Al is decreased when dipped in HNO3 due to

a) Formation of Oxide layer.
b) Formation of Hydroxide layer.
c) Formation of Hydride layer.
d) Reactivity never decreases.

29) Which metal is lighter than water

a) Hydrogen.
b) Lithium.
c) Sodium
d) Osmium

30) Ionic compounds are

a) Brittle.
b) Ductile.
c) Malleable
d) Amorphous

31) Metals found in the middle of reactivity series are generally found in following forms

a) Carbonates, chlorides and nitrides.
b) Carbonates, sulphides and nitrides.
c) Carbonates, Oxides and sulphides.
d) Chlorides, Oxides and sulphides.

32) What are the ions are presents in Na_2O

a) Na^{++}, O^{2-}
b) Na^+, O^{2-}
c) Na^{++}, O^-
d) Na^{++}, O^{2-}

33) During calcinations, following poisonous gas mainly produced.

a) CO
b) CO_2
c) SO_2
d) N_2O

34) Roasting is the method of producing

a) Sulphide to carbonate.
b) Sulphide to Oxide.
c) Oxide to Sulphide.
d) Sulphide to chloride.

35) The impurities that contaminated ores of metal?

a) Gravel
b) Gangue.
c) Garbage.
d) Gravel.

36) Most of the metal color is

a) Blue.
b) Grey.
c) White.
d) Golden.

37) Which of the following metal generally found in free on nature?

a) Ca
b) Cu
c) Au
d) Sn

38) During roasting following poisonous gas is mainly produced

a) CO
b) CO_2
c) SO_2
d) N_2O

39) Which of the following non-metal is liquid

a) Iodine.
b) Bromine.
c) Carbon.
d) Sulpher.

40) During smelting additional substance is added which combine with impurities which formed a fusible product and its name is

a) Slug.
b) Gangue.
c) Mud.
d) Flux.

41) When Which of the following metal produce amphoteric oxide

a) Fe.
b) Mn.
c) Al.
d) Zn.

42) Which of the following metal does not react with cold as well as hot water?

a) Na.
b) Mg.
c) Mn.
d) Fe.

43) The most abundant metal in earth crust is

a) Aluminum.
b) Iron.
c) Calcium.
d) Sodium.

44) Which of the following property is not related to Ionic compounds?

a) They are highly soluble in water.
b) They conduct electricity in solid state.
c) They have high melting point temperature.
d) They are brittle.

45) Second most abundant metal in earth crust is

a) Aluminum.
b) Iron.
c) Calcium.
d) Sodium.

46) Hematite is an ore of

a) Hg.
b) Fe.
c) Ca.
d) Na.

47) Which chemical process used to get metal on top in reactivity series?

a) Calcination.
b) Electrolysis.
c) Roasting.

d) Anodization.

48) Which is correct order as per reactivity of metal

1. Na > Mg > Zn > Cu.
2. Mg > K > Fe > Cu.
3. Na > Al > Fe > Cu.
4. Al > Na > Zn > Cu.

a) 2 & 3 are correct.
b) 1 & 3 are correct.
c) 2 & 2 are correct.
d) 2 & 4 are correct.

49) Between Ore and Mineral in which meat concentration is more

a) Ore
b) Mineral.
c) Both having the same concentration of metal and they are a synonym to each other.
d) A mineral having more concentration of metal than ore.

50) Rusting of iron could take place in

a) Distilled water.
b) Ordinary water.
c) Distilled and ordinary water.
d) None of the above.

51) Rose gold is an alloy of

a) Gold & Iron.
b) Gold & Copper.
c) Gold & Silver.
d) Gold & Zinc.

52) Molecular ammonia has

a) Only single bonds.
b) Only double bonds.
c) One single and 2 double bonds.
d) Only triple bonds.

53) Number of anions present in $MgCl_2$ is

a) 1
b) 2
c) 3
d) Not derive any anion.

54) Solder is an alloy of

a) Lead & tin.
b) Lead & zinc.
c) Zinc & tin.
d) Silver & Mercury.

55) Cinnabar is

a) Mercury oxide.
b) Mercury sulphide.
c) Mercury chloride.
d) Mercury nitrate.

56) Which of the following is not present in the form of the molecule in a compound

a) NaCl
b) C_2H_4
c) Fe_3O_4
d) H_2O

57) Which of the following is most reactive Element?

a) Ag
b) Cu
c) Au
d) Hg

58) The following metal could not be cut by a knife

a) Ca
b) Na
c) K
d) Li

59) When a mixture of copper oxide and copper sulphide heated

a) Copper sulphate is formed.
b) Copper is formed
c) Sulpher trioxide is formed
d) No reaction takes place

60) The hardest allotrope of carbon is

a) Graphite.
b) Carbon black
c) Diamond.
d) Fullerene

61) Which metal used for galvanization?

a) Zinc.
b) Copper.
c) Iron.
d) Manganese.

62) In a general number of electrons are present in the outermost shell of a metal atom is?

a) 2 & 3.
b) 2 to 4.
c) 1 to 3.
d) 5 to 8.

63) To refine copper by electrolysis the impure copper metal should be used as

a) Cathode.
b) Anode.
c) Either anode or cathode.
d) Electrolysis method could not be used for refining copper.

64) Which one of the following is wrong regarding bronze.

a) It's a good conductor of electricity.
b) It is an alloy of Copper and tin.
c) It is resistance to corrosion.
d) None of the above is wrong.

65) Which of the following metal not added to reduce Carat rating of gold

a) Cu.
b) Ag.
c) B
d) None of the above.

66) After electrolysis, the required metal is deposited at

a) Cathode
b) Anode.
c) The terminal type depends on the metal.
d) Deposited at the bottom of the container irrespective to the terminal.

67) Sodium metal is obtained by electrolysis of

a) Brine solution.
b) Sodium sulphate.
c) Sodium bi-carbonate.
d) Sodium oxide.

68) Aluminum metal is obtained by electrolysis of

a) Bauxite.
b) Aluminum chloride.
c) Alumina
d) Aluminum nitrate.

69) Most carbon element is stored in

a) Earth crust.
b) Atmosphere.
c) Ocean bed.
d) River bed.

70) Which of the following statement regarding non-meals is/are wrong?

1) They produce hydride when reacts with hydrogen.
2) They generally produce acidic or neutral oxides.
3) They are found mostly in solid state.
4) They are able to displace hydrogen from dilute acids only.

a) 2 only
b) 3 & 4 only
c) 1 &3 only
d) 4 only

71) Cinnabar to Hg production is related to the following processes

1) Calcination.
2) Roasting.
3) Electrolysis.

4) Refining.

a) 1 only
b) 2 only
c) 3 & 4 only
d) 2 & 3 only

72) Nonmetal oxides are

a) Acidic
b) Alkaline
c) Neutral
d) Either (a) or (c)

73) An element reacts with oxygen to give a compound with a high melting point. This compound is also soluble in water. What is that element?

a) C.
b) Si.
c) Ca.
d) Fe

74) Which of the following process is costliest

a) Electrolysis.
b) Roasting.
c) Calcinations.
d) Oxidation.

75) Which of the following metals are required to store in kerosene

1. Lithium
2. Potassium.
3. Sodium.
4. Calcium.

a) 1 & 2 only.
b) 1, 2 & 3 only.
c) 2 & 3 only.
d) 3 only.

76) Pure iron make hard by mixing with

a) Hydrogen.
b) Carbon.
c) Manganese.
d) Magnesium.

77) Nascent hydrogen atomicity is

a) 1
b) 2
c) 3
d) None

78) The alloy is an example of

a) Colloid.
b) A heterogenous mixture of metals.
c) The homogenous mixture of metals or metals & non metals.
d) A homogeneous mixture of two nonmetals.

79) Copper utensils become green when not used long days due to the formation of

a) Copper carbonate.
b) Copper oxide.
c) Copper Nitrate.
d) Copper shulphide.

80) Pure iron make hard by mixing carbon upto _____ %

a) 2
b) 4.5
c) 0.05
d) 1

[Answers & Explanations]

1) **(c).** Ductility is the ability to drawn into thin wire.

2) **(d).** Aqua regia is the strongest mixture of acids.it are molar mixture HNO_3 and H_2SO_4 in the ratio of 1:3. This mixture could dissolve Au.

3) **(c).** Bronze composition is 88% copper (metal) and 12 & tin (non-metal) and none have any nonmetal in their composition. Silver amalgam: Ag + Hg. Brass: Cu + Zn. Stainless steel: Fe + Cr + Ni + Mn.

4) **(a).** Galvanization is the method of coating Zinc on iron.

5) **(a).** Due to high oxidation property

6) **(c).** Gallium is the only metal which melts at 38 deg C.

7) **(b).** Aqua regia is the strongest mixture of acids.it are molar mixture HNO_3 and H_2SO_4 in the ratio of 1:3. This mixture could dissolve Au.

8) **(a).**

9) **(c).** Iodine is the lustrous nonmetal. It's an exception.

10) **(a).** Na & K readily react with oxygen and catch fire.

11) **(d).** Ductility is the ability to drawn into thin wire.

12) **(b).** Silver sulphide color is black.

13) **(b).** Hematite is Fe_2O_3.

14) **(a).**

15) **(b).** Zinc sulphide is also known as Zinc blende.

16) **(b).** It reduces Ferric oxide to iron by removing oxygen.
$Fe_2O_3 + 2Al = 2Fe + Al_2O_3$

17) **(c).** Aqua regia is the strongest mixture of acids.it are molar mixture HNO_3 and H_2SO_4 in the ratio of 1:3. This mixture could dissolve Au.

18) **(c).**

19) **(b).** MnO_2 is reduced to Mn by losing oxygen.

$3MnO_2 + 4Al = 3Mn + Al_2O_3$

20) **(a).** Thermite method is used to weld railways and repairing ships. It's generally used where movement of a broken component up to the maintenance shop is not possible.

21) **(b).**

22) **(b).** Due to the formation of oxide layer reactivity decreases.

23) **(b).** malleability is the property of the metal by means of which it could easily convert to thin metal.

24) **(d).** Amalgam is the alloy of mercury.

25) **(a).** Zinc sulphide is known as Zinc blend.

26) **(a).** HCl gas and CCl4 are not an ionic compound.

27) **(a).** Graphite is a non-metal which conduct electricity.

28) **(a).** Due to the formation of oxide layer Al activity decreases in nitric acid.

29) **(b).** Hydrogen is the lightest element. Lithium is lighter than water of density 0.543 kg/CuM. Sodium is about 1000 times denser than water. Osmium is the naturally found heaviest element.

30) **(a).** Because of the strong force

of attraction Ionic compounds are brittle.

31) (c).

32) (b). Na　O
　　　　　　1　　2

On ionization, it will produce 2 Na+ and one O^{2-} ion.

33) (b). calcination carried out for carbonate ores of metal. On heating, they produce an oxide of metal and CO_2.

$ZnCO_3 = ZnO + CO_2$

34) (b). Roasting is done for sulphide in presence of oxygen during heating with oxygen it produces SO_2.

$2ZnS + 3O_2 = 2ZnO + 2SO_2$

35) (b).

36) (b).

37) (c). Pt, Pd, Au are called Nobel metals.

38) (c). Roasting is done for sulphide in presence of oxygen During heating with oxygen it produces SO_2.

$2ZnS + 3O_2 = 2ZnO + 2SO_2$

39) (b). Iodine: Solid. Bromine: Liquid. Carbon: Solid. Sulpher: Solid

40) (a)

41) (c) These oxides react with acid and alkaline both. $Al2O3$ is an amphoteric oxide.

42) (d) Lanthanides are started from cerium (Ce) of atomic number 58.

43) (a) First most abundant metal is aluminum & second one is Fe.

44) (b) Due to lack of free electrons, ionic compounds are not able to conduct electricity in solid state. There bond strength is very high. In water, solution salts are ionized and conduct electricity.

45) (b) First most abundant metal is aluminum & second one is Fe.

46) (b) Hematite(Fe_2O_3)

47) (b) All metals at the top of the reactivity series extracted by electrolysis method. Such metals oxide cannot be deoxidized by simply heating with carbon.

48) (b).

49) (a). The minerals, which could use to extract metal proficiently, then called ore. All ore are minerals but not all minerals are ore.

50) (b). Distilled water not contains dissolved oxygen so it is unable to do rust.

51) (b).

52) (a).

53) (b). $MgCl_2 = Mg^{2+} + Cl^{2-}$

54) (a).

55) (b). Cinnabar is HgS

56) (a) NaCl does not form actual bonding but stays as a matrix.

57) (b) All other given elements are positioned below the copper element in the reactivity series.

58) (a).

59) (b). $2Cu_2O + Cu_2S \Rightarrow 6Cu + SO_2$

60) (c) Diamond is the hardest natural element.

61) (a)

[A COMPLETE GUIDE TO MCQ]

CHAPTER: 03

62) (c).

63) (b).

64) (a). Bronze is a bad conductor of electricity and all other statements are correct.

65) (c) Borax used for purifying gold. Copper & silver used to change carat rating of gold.

66) (a) The impure metal used as anode and metal deposited at the cathode during electrolysis purification as metal ions take electron at the cathode and then deposits here.

67) (a) NaCl is used for the production of Na and known as Brine.

68) (c) Al_2O_3 also known as Alumina

69) (C) Most carbon is deposited in ocean bed due to rain after cooling of the earth crust.

70) (d) All non-metals are located below the hydrogen in the reactivity series.

71) (b) Cinnabar is HgS Mercury sulphide. HgS on the presence of oxygen due to heating is formed mercury oxide and on further heating it becomes Hg. The first process is called roasting.

$$2HgS + 3O_2 = 2HgO + 2SO_2$$
$$2HgO = 2Hg + O_2$$

72) (d) CO_2 is a nonmetallic oxide and acidic in nature whereas H_2O is nonmetallic oxide but neutral in nature.

73) (c). SiO_2, Fe_3O_4 is insoluble in water. CaO & CO_2 are soluble in water but CaO has a high melting point as it is solid.

74) (a) Electrolysis is the costliest method to get metal in pure form.

75) (b) Except Ca all other given elements react with water and oxygen vigorously so they are stored in kerosene.

76) (b) Carbon makes pure iron hard and 0.05% C mixes with iron to make it hard.

77) (a) Nascent hydrogen is the atomic form of hydrogen, which just produced during the chemical reaction.

78) (c)

79) (a) Copper reacts with CO_2 in presence of moisture and form Copper carbonate that is a green color.

80) (c) To make hard steel 0.03 to 0.05% C is mixed. Up to 4.6% C could mix with iron, which gives cast iron.

Carbon & Its Compounds

CHAPTER: 04

1) The bond between two identical nonmetallic atoms has a pair of electrons

a) Are equally shared between two atoms.
b) Transferred completely from one atom to another.
c) With identical spins.
d) Equally shared between them.

2) Which of the followings are correct: During welding Ethyne burned with oxygen other than simple air because

1. Burning with oxygen produces more heat.
2. Burning with air produces soots.
3. Burning with air produces heavy sounds.

a) 1 only.
b) 2 & 3 are correct.
c) 1 & 2 are correct.
d) 1 & 3 are correct.

3) Margarine are produced by?

a) Hydrogenation.
b) Substitution reaction.
c) Polymerization.
d) Coagulation.

4) Which of the following compound having a cyclic form with the presence of double bonds

a) Cyclohexane.
b) Benzene.
c) Hexa-chloro benzene.
d) Both (b) & (C).

5) Which of the following compound should have a specific scent

a) Cyclohexane.
b) Benzene.
c) Gammexane.
d) All of the above.

6) When sodium carbonate is added with Methanoic acid

a) CO_2 gas evolved.
b) CO evolved.
c) HCN gas evolved.
d) H_2 gas evolved.

7) SCUM is

a) Precipitation of calcium salt of the organic fatty acid.
b) Slury of the magnesium salt of the organic fatty acid.
c) Precipiation of calcium & magnesium salt of the organic fatty acid.
d) Slury of calcium & Potassium salt of the organic fatty acid.

8) Which of the followings contain triple bonds

1. Nitrogen.
2. Acetylene.
3. Ethane.
4. Propene

a) 1 & 4 are correct.
b) 2 & 3 are correct.
c) 1 & 2 are correct.
d) 2 & 3 are correct.

9) Which of the following is biodegradable?

a) Soap.
b) Detergent.
c) Both (a) & (b)
d) None.

10) When ethanol is oxidized with alkaline potassium permanganate it produces

a) Ethanoic acid.
b) Ethane.
c) Ethyne.
d) Methanoic acid.

11) Which allotrope of carbon used as a lubricant?

a) Charcoal.
b) Graphite.
c) Carbon nono-tube.
d) Fullerene.

12) Halogenations reaction is not possible for

a) Alkanes.
b) Alkynes.
c) Alkenes.
d) None of the above.

13) IUPAC name of the following compound is

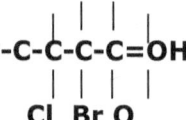

a) 1-Chloro 3-bromo propanoic acid.
b) 2- bromo 4-Chloro Butanoic acid.
c) 1-Chloro 3-bromo butanoic acid.
d) 2- bromo 4-Chloro Propanoic acid.

14) Which one is not correct

a) Methanol is lethal for health.
b) Ethanoic acid is highly inflammable.
c) Ether should keep in dark glass container.
d) Mallic acid is apple cider vinegar.

15) For good health, we should consume

a) Saturated animal fat.
b) Unsaturated fat.
c) Both (a) and (b).
d) Mainly unsaturated plant fat.

16) Formalin which is used to preserve biological samples is

a) A solution of formic acid and water.
b) A solution of formaldehyde.
c) A solution of foam.
d) A solution of NaCl and Formaldehyde.

17) For welding following gas is used

a) Ethylene.
b) Ethyne.
c) Propyne.
d) Methane.

18) Following compound is used as antifreeze

a) Glycerin.
b) Ethanol.
c) Salt.
d) All of the above.

19) Unsaturated hydrocarbons burn with

a) Red flame.
b) Yellow flame.
c) Blue flame.
d) Orange flame.

20) Which part of the flame have the highest temperature

a) Transparent part.
b) Orange part.
c) Yellow part.
d) Blue part.

21) Major constituents of CNG is

a) Methane.
b) Ethane.
c) Propane.
d) Butane.

22) Which of the following is different from others.

a) Ethylene.
b) Benzene.
c) Propane.
d) Ethene.

23) Carbon Nano tube is an

a) An isotope of carbon.
b) An allotrope of carbon.
c) An isomer of Fullerene.
d) Hardest form of carbon.

24) Most clean fuel is

a) LPG
b) Kerosene.
c) CNG
d) Aviation petrol.

25) The color of aviation petrol is

a) White.
b) Red.
c) Blue.
d) Colorless.

26) Soap molecules are mainly salt of

a) Sodium & Calcium.
b) Magnesium & Potassium.
c) Potassium & lithium.
d) Sodium & Potassium.

27) The shape of carbon allotrope carbon Nano tube is

a) Spherical.
b) Geodesic.
c) Cylindrical.
d) Tetrahedron.

28) The general formula of alkynes

a) C_nH_{2n}
b) C_nH_n
c) C_nH_{2n+1}
d) C_nH_{2n+2}

29) The general formula of alkanes

a) C_nH_{2n}
b) C_nH_n
c) C_nH_{2n+1}
d) C_nH_{2n+2}

30) The general formula of alkenes

a) C_nH_{2n}
b) C_nH_n
c) C_nH_{2n+1}
d) C_nH_{2n+2}

31) The tail part of soap molecules are

a) Positively charged hydrocarbon.
b) Negatively charged hydrocarbon.
c) Non-polar hydrocarbon.
d) None of the above.

32) The tail part of soap molecules are

a) Hydrophobic.
b) Hydrophilic.
c) Hygroscopic.
d) Sublimate able.

33) The head part of soap molecules are

a) Hydrophobic.
b) Hydrophilic.
c) Hygroscopic.
d) Sublimate able.

34) Which of the following will

decolorize bromine water(aq)

a) Hydrogenated oil.
b) Cooking oil.
c) Butter.
d) All of the above.

35) In homologous series

1. All compounds have a same general formula.
2. Have same physical properties.
3. Have same chemical properties.
4. Have atomic mass difference 14 for two successive compounds.

a) 1,3 & 4 are correct.
b) 1,2 & 3 are correct.
c) 1 & 4 are correct.
d) 2 & 4 are correct.

36) According to increasing order of stability

a) Alkanes < Alkenes < Alkynes.
b) Alkenes < Alkanes < Alkynes.
c) Alkenes < Alkynes < Alkanes
d) Alkynes < Alkenes < Alkanes.

37) Pyrolysis is

a) Decomposition of the compound on heating in absence of air.
b) Decomposition of the compound on heating in presence of excess air.
c) Breaking of the large organic compound.
d) Opposite of polymerization.

38) Soots are produced during burning of

a) Alkanes.
b) Alkynes.
c) Alkenes.
d) Both (b) & (C)

39) Breaking is

a) Decomposition of the compound on heating in absence of air.
b) Decomposition of the compound on heating in presence of excess air.
c) Breaking of the large organic compound.
d) Opposite of Pyrolysis.

40) In Micelle

1) Hydrophobic tail in inside the sphere.
2) Hydrophilic tail in inside the sphere.
3) The hydrophobic head is inside the sphere.
4) Hydrophilic head is on the curved surface of the sphere.

a) 1 & 4 are correct.
b) 2 & 3 are correct.
c) 1 is correct.
d) 3 is correct.

41) Which metals is/are used as a catalyst during the hydrogenation process

a) Pt
b) Pt & Pd
c) Ni & Pd
d) Ni

42) Saturated hydrocarbons are burned with

a) Red flame.
b) Yellow flame.
c) Blue flame.
d) Orange flame.

43) In which component have weakest Carbon Carbon bond.

a) Alkynes.
b) Alkanes.
c) Alkaline.
d) Alkenes.

44) $CH_3-CH_2OH \xrightarrow{Conc. H_2SO_4}$

a) Ethelene.
b) Ethane.
c) Methane.
d) Ethene.

45) The main constituent of Denatured sprit is

a) Methanol.
b) Ethanol.
c) Butanol.
d) Isoleucine.

46) Sprit is denatured by

a) Methanol.
b) Ethanol.
c) Butanol.
d) Chloroform.

47) Which of the following will undergo an addition reaction

a) CH_4
b) C_2H_6
c) C_3H_8
d) C_2H_2

48) Gasol is

a) Petrol mixed with alcohol at any percentage.
b) Petroleum mixed with alcohol up to 20%.
c) Petrol mixed with alcohol up to 20%.
d) Petrol mixed with alcohol more than 20%.

49) Which component is not used to denature spirit

a) Blue vitriol.
b) Lysine.
c) Pyridines.
d) Butanol

50) $CH_3-COOC_2H_5 \xrightarrow{NaOH} C_2H_5OH +$

a) Acetic acid.
b) Formic acid.
c) Methane.
d) Ethylene.

51) Glacial acid is the old name of

a) Formic acid.
b) Tartaric acid.
c) Acetic acid.
d) None of the above.

52) Vinegar is

a) 5 to 8% Acetic acid in Water.
b) 10 to 20% Acetic acid in Water.
c) 5 to 8% Acetic acid in ethanol.
d) 30 to 40% Malic acid in Water.

53) Two show isomerism property one hydrocarbon should have minimum_____ number of carbon atoms

a) 2
b) 3
c) 4
d) 5

54) Isomers have

a) Same molecular formula and different atomic mass.
b) Same molecular formula but different atomic structure.
c) Different molecular formula but same atomic structure.
d) Different molecular formula but same atomic weight.

55) Alkanes are

1. Saturated hydrocarbon.
2. Unsaturated aldehyde.
3. Minimum one double bond present.
4. Take part in hydrogenation.

a) 1 only.
b) 2 & 3.
c) 1,3 & 4.
d) 2 & 4.

56) Which one is the functional group of an aldehyde

a) –COOH
b) –CHO
c) -CO
d) =C

57) Which one is the functional group of carboxylic acid

a) –COOH

b) –CHO
c) -CO
d) =C

58) Which one is the functional group of a ketone?

a) –COOH
b) –CHO
c) -CO
d) =C

59) Isometric pair is

a) Ethane and propane.
b) Butane and 2-methyl propane.
c) Butane and propyne.
d) Ethane and ethylene.

60) Detergents are sodium salts of a long chain of

a) Ketone.
b) Aldehyde.
c) Sulphonic acids.
d) Carboxylic acids.

61) Which allotrope of carbon having a hexagonal shape in layer form

a) Graphite.
b) Carbon nano tube.
c) Fullerene.
d) Diamond.

62) The difference of atomic mass between two successive compounds in homologous series is

a) 12
b) 14
c) 15
d) 16

63) Successive members of homologous series differ by

a) –CH$_2$ unit.
b) –CH$_3$ unit.
c) –CH unit.
d) None of the above

64) In homologous series a successive member of C$_5$H$_{10}$ is

a) C$_4$H$_8$.
b) C$_6$H$_{12}$.
c) C$_5$H$_{12}$.
d) C$_4$H$_{12}$.

65) In homologous series compound before C$_3$H$_7$OH is

a) C$_4$H$_8$OH
b) C$_2$H$_5$OH
c) C$_4$H$_9$OH
d) C$_6$H$_9$OH

66) Which of the following element does not have allotrope

a) Oxygen.
b) Sulpher.
c) Phosphorous.
d) Zinc

67) Covalent bond is a bad conductor of electricity due to

1. Unable to produce ion in aqueous solution.
2. Have higher bond strength than an ionic bond.
3. Complete shift of electron is not possible.

a) 1 only.
b) 1 & 2.
c) 1 & 3.
d) 2 & 3.

68) In which elements molecule triple bond is found

a) Oxygen.
b) Nitrogen.
c) Bromine.
d) Sulfur

69) Versatile nature of carbon is observed due to

a) Catenation property.

Carbon & its Compounds MCQ (class X)

b) Tetra valency
c) Both (a) & (b).
d) None of the above.

70) Ethanol is not used for the preparation of

1) Drinks.
2) Wood polish.
3) Cough syrup.
4) Paints.

a) 1 only
b) 2 only
c) 1 & 4 only
d) None of the above.

71) For sterilizing wounds following alcohol is used

a) Ethyl alcohol.
b) Methyl alcohol.
c) Acetic acid.
d) Butanol.

72) Catenation property found in

a) Si.
b) C.
c) S
d) All of the above

73) Geodesic shape found in following allotrope of carbon?

a) Fullerene.
b) Diamond.
c) Graphite.
d) Carbon nanotube.

74) What is the color of flame when oxygen supply is sufficient for burning

a) Orange.
b) Red.
c) Yellow.
d) Blue.

75) Which alcohol is mixed with petrol

1. Butanol.
2. Propanol.
3. Ethanol.

a) 1 & 2 only.
b) 2 & 3 only.
c) 3 only.
d) All of the above.

76) Ethanol reaction with Ethanoic acid when it is heated with conc. Sulphuric acid is called

a) Esterification.
b) Emulsification.
c) Saponification.
d) Iodoform reaction.

77) Which of the following types of reaction is common form organic compounds

a) Combination reaction.
b) Addition reaction
c) Substitution reaction.
d) All of the above.

78) The leak of LPG is detected by the smell of

a) Butane.
b) Ethyl mercaptan.
c) C S Gas.
d) Methane

79) Main constituents of LPG (Liquefied Petroleum Gas) is

a) Methane.
b) Propane.
c) Butane.
d) Ethane.

80) IUPAC name of vinegar is

a) Ethanoic acid.
b) Methanoic acid.
c) Dihydroxybutanoic acid.
d) Hydroxybutanoic acid.

[A COMPLETE GUIDE TO MCQ]

CHAPTER: 04

[Answers & Explanations]

1) (a).

2) (c). Excessive oxygen used to produce more heat and no soot.

3) (a). Its imitation butter made form unsaturated plant oil by hydrogenation method.

4) (d). Benzene:

Cyclohexane:

Hexachloro benzene:

5) (d). All are aromatic compounds and have a specific smell.

6) (a).

7) (c). SCUM is a layer of dirt or froth on the surface of a liquid.

8) (c). Ethane has all single bond and Propene have a double and single bond.

9) (a).

10) (a).

11) (b). Due to layer crystal structure graphite used as a lubricant.

12) (d). Both Saturated and unsaturated aliphatic compounds take part in halogenations reactions.

13) (b). Functional group carbon atom serial number should be lowest.

14) (b).

15) (d).

16) (b). it is 10% solution of formaldehyde.

17) (b). C_2H_2

18) (d).

19) (b). Due to the presence of less hydrogen unsaturated hydrocarbons are burn with a yellow flame.

20) (d).

21) (a). Methane is the main constituents but also contain some amount of propane and ethane.

22) (b). Benzene is aromatic compound and all others are aliphatic compound.

23) (b).

24) (c). CNG (mainly contain CH_4) after burning produces CO_2 and water vapor only.

25) (a). Originally it is white in color but sometime blue or green color dye is mixed.

26) (d).

27) (c). Cylindrical. Geodesic (hemispherical dome) for Fullerene and Tetrahedron for diamond.

28) (c).

29) (d).

30) (a).

31) (c).

32) (a).

Carbon & its Compounds MCQ (class X)

33) **(b)**.

34) **(b)**. Cooking oil is unsaturated oil and it will react with bromine by means of halogenation reaction.

35) **(a)**. Homologous compounds have different physical properties.

36) **(d)**. Stability decrease from single bond to triple bonds.

37) **(a)**.

38) **(d)**. Saturated hydrocarbons does not produce soot.

39) **(c)**.

40) **(a)**.

41) **(c)**.

42) **(c)**. Due to complete combustion.

43) **(a)**. Alkynes have a triple bond and it is weakest among single and double bonds. Alkanes and alkenes have single and double bond respectively.

44) **(d)**.

45) **(b)**.

46) **(a)**.

47) **(d)**. Unsaturated hydrocarbons only take part in an addition reaction. Option a to c are saturated hydrocarbon as they have general formula C_nH_{2n+2}

48) **(c). Gasol** is a mixture of alcohol (up to 20% V/V) and petrol.

49) **(d)**. Blue vitriol ($CuSO_4, 5H_2O$), Lysine, Pyridines are mixed with ethyl alcohol to make denatured sprit.

50) **(a)**.

51) **(c)**. Acetic acid looks glacier so it's named like this.

52) **(a)**. Malic acid form apple cider vinegar.

53) **(c)**.

54) **(b)**. Isomers are two or more compounds with the same formula but a different arrangement of atoms in the molecule and different properties.

55) **(a)**.

56) **(b)**.

57) **(a)**.

58) **(c)**.

59) **(b) Butane:**

H H H H
| | | |
H—C—C—C—C—H
| | | |
H H H H

2-methyl propane:

60) **(c)**.

61) **(a)**. The shape of carbon Nano tube is right angle cylinder, for fullerene it is geodesic (hemispherical dome) and for diamond is a tetrahedron.

62) **(b)**. The atomic mass difference is 14 as atomic mass of $-CH_2$ is 12 +2×1=14 amu.

63) **(a)**.

64) **(b)**. Homologous series differ $-CH_2-$

65) (b). Homologous series differ –CH_2–

66) (d). An allotrope of Oxygen is Ozone, Sulpher allotrope is rhombic and monoclinic, Phosphorous allotropes are red, white, black, and violet phosphorous.

67) (c). Ionic bond strength is more than a covalent bond.

68) (b).

69) (c). Tetra valency and catenation property is the main cause behind versatile nature of carbon.

70) (d). All the substances use ethyl alcohol.

71) (a). In the form of tincture iodine.

72) (d). All the given elements have catenation property.

73) (a). Buckminster Fullerene is an allotrope of shape geodesic or hemispherical dome.

74) (d). When oxygen supply is sufficient all hydrogen atom get required oxygen so its color is blue and the temperature is maximum.

75) (c). Ethanol i.e. ethyl alcohol is mixed with petrol up to 20%. This is called blended petrol or gasol.

76) (a). Ester is an organic salt. When an organic acid reacts with alcohol, it produces the ester.

77) (d).

78) (b). Ethyl Mercaptan is a colorless or yellowish liquid or a gas with a pungent, garlic or skunk-like odor. Ethyl marcaptan could detect if its present 1-160ppm in the air.

79) (c). LPG gas contains propane, propylene, butane, and butylene. Main constituent is **Butane**.

80) (a). Vinegar is acetic acid i.e. CH_3COOH. It contains 2-carbon atoms and one carboxylic group. Therefore, its IUPAC name is Ethanoic Acid.

Periodic Classification Of Elements

1) Dobereiner placed how many elements in a group

a) 3 elements.
b) 4 elements.
c) 6 elements.
d) 8 elements.

2) The last element which follows Octave rule is

a) Ga.
b) Ca.
c) Sc.
d) Ge.

3) Which of the following is not an example of Dobereiner triads

a) Cl, Br, I.
b) Li, Na, K.
c) Ca, Sr, Ba.
d) Be, Ga, Ca.

4) According to Mendeleev's periodic law elements are arranged in the periodic table in the order of

a) Increasing atomic number.
b) Increasing atomic masses.
c) Increasing atomic weight.
d) Increasing mass number.

5) Dobereiner identified how many elements in his table of triads

a) 3 elements.
b) 6 elements.
c) 9 elements.
d) 12 elements.

6) In the Modern periodic table, elements are arranged in the periodic table in the order of

a) Increasing atomic number.
b) Increasing atomic masses.
c) Increasing atomic weight.
d) Increasing mass number.

7) Mendeleev started his work for classification of elements with

a) 56 elements.
b) 63 elements.
c) 92 elements.
d) 65 elements.

8) According to Dobereiner Triads

a) An atomic number of middle element was roughly the average of the atomic numbers of the other two elements.
b) Atomic mass of middle element was roughly the average of the atomic masses of the other two elements.
c) The atomic weight of a middle element was roughly the average of the atomic weights of the other two elements.

d) A mass number of middle element was roughly the average of the atomic mass numbers of the other two elements.

9) Which of the following element was not predicted by Mendeleev

a) Gallium.
b) Beryllium.
c) Scandium.
d) Germanium.

10) Octave law of periodic classification of element stated by

a) Mendeleev.
b) Mendel.
c) Newland.
d) Dobereiner.

11) Which of the following statement(s) about modern periodic table are incorrect?

1) Elements are arranged on the basis of increasing atomic number.
2) Elements isotopes are placed adjoining group(s) in the periodic table.
3) Radioactivity is a periodic property.
4) Elements are arranged based on increasing atomic masses.
5) Hydrogen placed out of the table.

a) 1 & 2 are incorrect.
b) 3 & 4 are incorrect.
c) 2, 3 & 4 are incorrect.
d) 2 & 5 are incorrect.

12) Octave rule classified how many elements

a) 52 elements.
b) 54 elements.
c) 56 elements.
d) 66 elements.

13) Which of the following is the outermost shell for the element of period 3

a) K shell.
b) L shell.
c) M shell.
d) N shell.

14) A heaviest natural element in the modern periodic table is

a) Os.
b) Sm.
c) Fr.
d) Nb.

15) Nobel gases are placed in the separate group due to

1) Similar inert behavior.
2) Similar electronic configuration.
3) Same valance.
4) Available in low concentration.

a) 1 & 2 are correct.
b) 3 & 4 are correct.
c) 2, 3 & 4 are correct.
d) All are correct.

16) An element X having atomic number 12 combines with Nitrate ion, the formula of that compound will be

a) XNO_3.
b) $X(NO_3)_2$.
c) X_2NO_3.
d) $X_3(NO_3)_2$.

17) Second group elements are called

a) Alkali metals.
b) Alkaline earth metals.
c) Halogen elements.
d) Chalcogen elements.

18) Which group elements are called Chalcogen

a) Group-15.
b) Group-16.
c) Group-17.
d) Group-18.

19) Following groups are called "transition element" group

a) First and second group.
b) Third to the twelfth group.
c) Thirteen to eighteen group.
d) Fifth and sixth group.

20) Which one is odd?

a) Calcium.
b) Strontium.
c) Magnesium.
d) Potassium.

21) Which of the following element is not comes under "*chalcogen*" family but present in the same group?

a) Sulfur.
b) Oxygen.
c) Selenium.
d) Polonium.

22) The third period contains 8 elements, out of these elements how many elements are metals?

a) 5.
b) 3.
c) 4.
d) 2.

23) *Last Nobel gas in the periodic table is*

a) Rn.
b) At.
c) Xe.
d) Ne.

24) Which of the following element does not have electron equal to Ar?

a) K+
b) Cl-
c) Ca++
d) Na+.

25) Where would you locate the element with atomic number 12 in the modern periodic table

a) Period 3, Group 1.
b) Period 2, Group 1.
c) Period 3, Group 2.
d) Period 1, Group 2.

26) Which of the following blocks of elements comprises transition elements

a) S-Blocks elements.
b) p-Blocks elements
c) d-Blocks elements
d) f-Blocks elements

27) Down a group, the electron affinity will

a) Increases.
b) Decreases.
c) Remains same.
d) First, increase then decrease.

28) As on today longest period of the modern periodic table is

a) Fourth.
b) Sixth.
c) Eighth.
d) Seventh.

29) Which element(s) position is controversial in the modern periodic table?

a) Hydrogen.
b) Nobel elements.
c) Halogen elements.
d) Hydrogen, Lanthanides and actinides.

30) Highly reactive metals are present in

a) 1st group.
b) 2nd group.
c) 3rd group.
d) 4th group.

31) Which of the following gives the correct increasing order of atomic radius?

a) O>F>N.

b) F>O>N.
c) N>F>O.
d) N>O>F.

32) Mendeleev was predicted the existence of

a) 2 elements.
b) 3 elements.
c) 4 elements.
d) 5 elements.

33) Electronic configuration of Al^{3+} is

a) 2,8,3.
b) 2,8.
c) 2,8,6
d) 2,8,8

34) Arrange the following elements in the order of their increasing metallic character

a) Cl,Si,Al,Mg..
b) Al,Mg,Si,Cl.
c) Cl,Mg,Si,Al.
d) Mg,Al,C,Si

35) The electron affinity of an element depends on

a) Atomic size.
b) Nuclear charge.
c) Atomic number.
d) Atomic size and nuclear charge both.

36) Alkaline earth metals comprise the following group

a) Group-1.
b) Group-2.
c) Group-3.
d) Group-17.

37) An element having atomic number 24 will locate in

a) 4th period & 6th group.
b) 5th period & 6th group.
c) 6th period & 4th group.
d) 4th period & 2nd group.

38) Which of the following elements would lose an electron easily?

a) Mg.
b) Na.
c) K.
d) Ca.

39) Which of the following statement about electron affinity is wrong?

a) It causes energy to be absorbed.
b) It causes energy to be released.
c) Electron affinity expressed in volts.
d) It involves the formation of anions.

40) Which of the following set of elements is written in order of their increasing metallic character?

a) Be, Mg, Ca.
b) Na, Li, K.
c) Mg, Al, Si.
d) C, O, N.

41) Lanthanides start from atomic number

a) 56.
b) 57.
c) 58.
d) 90.

42) Which of the following factor affects ionization potential?

a) Electron affinity.
b) Atomic size.
c) Electron negativity.
d) Mass number.

43) Metalloids found in following groups?

a) 12, 13 & 14.
b) 11.12 & 13.
c) 2, 3 & 4.
d) 14, 15 & 16.

44) Mendeleev periodic table consists of

a) 7 periods and 12 groups.
b) 6 periods and 8 groups.
c) 6 periods and 18 groups.
d) 7 periods and 18 groups.

45) Ionization potential across a period from left to right will

a) Increases.
b) Decreases.
c) Remain same.
d) First, increase & then decrease.

46) What type of oxide would Eka-Aluminium form?

a) EO_3.
b) E_3O_2.
c) E_2O_3.
d) EO_2.

47) Lanthanides and Actinides are also called

a) Transition elements.
b) Nobel elements.
c) Inner transition elements.
d) Rare elements.

48) Which of the following elements will form acidic oxide?

a) An element of atomic number 3.
b) An element of atomic number 7.
c) An element of atomic number 12.
d) An element of atomic number 13.

49) Lithium belongs to

a) Alkali metals group.
b) Alkaline earth metals group.
c) Chalcogen family.
d) None of the above.

50) A least reactive element in halogen group is

a) Fluorine.
b) Chlorine.
c) Iodine.
d) Bromine.

51) Ionization potential across a group from top to bottom will

a) Increases.
b) Decreases.
c) Remain same.
d) First, increase & then decrease.

52) Eka-Boron which was predicted by Mendeleev is named after discover as

a) Thorium.
b) Gallium.
c) Germanium.
d) Scandium.

53) Lanthanides elements are also called

a) p-block elements.
b) d-block elements.
c) f-block elements.
d) Transition elements.

54) Inner transition elements all are -

a) Metals.
b) Metalloids.
c) Nobel metals.
d) Rare earth metals.

55) Which statement(s) are not correct regarding a metal?

1. All are a good conductor of electricity.
2. All are found in the solid form.
3. All form an acidic oxide.
4. All are having high melting point temperature than non metal.

a) 1, 3 & 4.
b) 2 & 3.
c) 2, 3 & 4.
d) 1, 3 & 4.

56) The only halogen which is a radioactive element in nature is

a) Fluorine.
b) Iodine.

c) Astatine.
d) None of the halogens is radioactive.

57) Iodine is solid due to

a) A large value of dipole moment.
b) Strong covalent bond.
c) Strong metallic bond.
d) A strong hydrogen bond is present.

58) On moving from left to right in the periodic table, the radioactivity

a) Decreases.
b) Increases.
c) First, increase and then decrease.
d) It is not a periodic property.

59) Modern periodic table has _____ blocks.

a) 3
b) 4.
c) 5.
d) 6.

60) Which of the following element would lose an electron easily?

a) Na.
b) Mg.
c) Ca.
d) K.

61) Which of the following is only solid halogen

a) Fluorine.
b) Bromine.
c) Iodine.
d) Astatine.

62) Who developed modern periodic table?

a) Mendeleev.
b) Bhor.
c) Rutherford.
d) Mosley.

63) Which of the following properties could not be predicted in the periodic table?

1. The nature of its oxides.
2. A number of isotopes present in nature.
3. Radioactivity.
4. The formula of its oxides.

a) 1, 3 & 4.
b) 2 & 3.
c) 2, 3 & 4.
d) 1, 3 & 4.

64) Which one is first alkali metal in the periodic table

a) Hydrogen.
b) Lithium.
c) Sodium.
d) Potassium.

65) Last alkali metal in the periodic table is

a) Cesium.
b) Thorium.
c) Actinium.
d) Francium.

66) Which period in the modern periodic table is called incomplete period?

a) 5th period.
b) 6th period.
c) 7th period.
d) None of the above.

67) Which of the following remains constant along the period from left to right?

a) Atomic radius.
b) Metallic property.
c) Valence.
d) None.

68) Which of the following always increases on going from top to bottom in a group?

a) Metallic character.
b) Electro negativity.
c) Oxidizing power.
d) Atomic radius.

69) Number of metalloids present in the modern periodic table is

a) 6 elements.
b) 5 elements.
c) 7 elements.
d) 8 elements.

70) Electro negativity of elements along a period from left to right is

a) Increases.
b) Decreases.
c) Remain constant.
d) First decrease and hen increase.

71) Last shell of the last element of any period

a) Electrons are filled completely.
b) Electrons are half filled.
c) Electrons are incomplete.
d) The single electron will present.

72) Which statement(s) are wrong regarding Mendeleev's periodic table?

1. All Nobel gases were present in the 8th group.
2. Groups are not subdivided into sub groups as a modern periodic table.
3. Isotopes are placed separately in a group.
4. Hydrogen is located in the first group.

a) 1, 3 & 4.
b) 2 & 3.
c) 2, 3 & 4.
d) 1, 2 & 3.

73) Which of the following remain constant when moving down a group

a) Atomic radius.
b) Metallic property.
c) Electron affinity.
d) Valence.

74) Catenation property found in following group elements?

a) Group-2.
b) Group-14.
c) Group-17.
d) Group-18.

75) Octave rule was valid for

a) Lighter elements only.
b) Heavier elements only.
c) Both lighter and heavier elements.
d) Only for gaseous elements.

76) Which family of elements outer most shell contains 7 elecrons.

a) Alkali metals.
b) Halogens.
c) Chalcogens.
d) Inner transition metals.

77) Three elements Be, As & P are

a) Metal, non-metal, metalloids respectively.
b) Non-metal, metal, metalloids respectively.
c) Metalloids, metal, non-metal, respectively.
d) Metal, metalloids, non-metal, respectively.

78) An element with electronic configuration 2,8,8 will locate in the modern periodic table in

a) Group 8.
b) Group 10.
c) Group 18.
d) Group 2.

79) Which of the following elements A, B, C & D with atomic number 3, 12, 20 & 22 respectively are in the same group?

a) A, C & D.
b) B & C

Periodic Classification of Elements MCQ (class X)

c) A & D.
d) B & D.

80) 18th Group of the periodic table have a valance

a) Zero.
b) Seven.
c) Eight.
d) Eighteen.

[Answers & Explanations]

1) **(a).** Dobereiner brought triads concept of element classification and he identified some groups having **three** elements each. So it's triads.

2) **(b).Calcium** was the last element on which Octave rule is applicable.

3) **(d).Gallium** was invented after Mendeleev and it was not known to Dobereiner.

4) **(b).** Mendeleev periodic law depends on increasing **Atomic masses** of the elements. The atomic number used for the modern periodic table. Atomic weight ant mass number never used for periodic classification of elements.

5) **(c).** Dobereiner identified only three triads so total element 3×3 =9 Elements

6) **(a).** Follow explanation of Q-4.

7) **(b).** At the time of Mendeleev, only **63** elements were known. **56** elements were known at the time of Newland.

8) **(b).** Dobereiner triads' concept depends on atomic weight.

9) **(b).** Mendeleev was predicted 3 elements Scandium, Gallium and Germanium. Beryllium was known to Mendeleev.

10) **(c).** Newland proposed Octave law of element classification.

11) **(c).** Except 1 and 5 all other statements are incorrect because elements isotopes are not placed separately, radioactivity is not a periodic function and Mendeleev periodic table built on increasing order of atomic masses.

12) **(c).** Octave rule classified 56 elements.

13) **(c).** 3rd period contains elements having an atomic number from 11 to 18. K=2, L=8, Total =2+8=10. So K & L shell will completely fill up before 3rd period. The electron will be filled up next shell M from 3rd period.

14) **(a).** Osmium (Os) is the heaviest element in the modern periodic table. Its density is 22.59 gm/CC.

15) **(d).** All statements are correct.

16) **(b).** Atomic number of X = 12. Electronic configuration K(2)L(8)M(2). So this element will lose 2 electrons from outer most shell to achieve Nobel gas configuration i.e. K(2)L(8). Valency of X is 2 and for NO_3 it is 1

Formula: $X(NO_3)_2$

17) **(b).** Alkaline earth materials occur in nature and artificially cannot be formed (till date). Elements are Be, Mg, Ca, Sr, Ba & Ra.

18) **(b).** 16th group elements are called chalcogen (O, S, Se & Te) except Po (Polonium).

19) **(b).** 3rd group to 12th group. But at present IUPAC 4th group to 11th group elements are called transition elements.

20) (d). Potassium is **Alkali metal** of group 1 where as all other elements (Ca, Mg, Sr) are **alkaline earth metals** of group 2.

21) (d). Except for Polonium all other elements are chalcogen.

22) (b). Out of 8 elements 3 are metals (Na, Mg, Al),1 metalloid(Si),4 are non-metals(P,S, Cl, Ar).

23) (b). Last Nobel gas in the periodic table is Astatine (At). It is radioactive.

24) (d). Na atomic number 11.So Na+ electronic configuration 2, 8, 1 and electronic configuration of Argon (atomic number 10) is 2, 8, 8.

25) (c). Atomic number=12. Electronic configuration 2, 8, 2 and shells are K, L, M. "n" for last shell number M=**3** and outer most shell electron number **2**. So element will be placed at the 3rd period,2nd Group. The element is Mg.

26) (c). The elements which have **incomplete d-sub shell** are called transition element. They are d-block elements.

27) (b). Electron affinity is the energy released when an electron is added to a neutral atom to from a negative ion. As down the group size of the atom is increased so energy released when an electron is added to a neutral atom from a negative ion will decrease due to less potential.

28) (b). As on today longest period of the modern periodic table is the Sixth period. It contains **32 elements** including 14 Lanthanides.

29) (d). Hydrogen has similarities with Halogen elements and alkali elements both but positioned at the 1st group. Similarly, at Lanthanide and Actinide 14 different elements are positioned at one place each.

30) (a). **1st group** elements outer shell electron easy to lose so they are highly reactive.

31) (d). Towards left, to the right of a period, atomic radius decreases due to increase in the nuclear positive charge for same outer sub shell.

32) (b). Eka-Boron (Scandium),Eka-Aluminum (gallium) & Eka-Silicon(Germanium).

33) (b). Aluminum atomic number Z=13.So electronic configuration is 2, 8, 3. But it lost 3 electrons of outer shell & become **Al^{3+}**. So electronic configuration will be **2,8(Neon)**

34) (a). Along a period metallic property increases from right to left.

35) (d). Electron affinity is the energy released when an electron is added to a neutral atom to from a negative ion. Increase in atomic size and nuclear charge both having an effect on released energy due to electron addition.

36) (b). Alkaline earth materials occur in nature and artificially cannot be formed (till date). Elements are Be, Mg, Ca, Sr, Ba & Ra.

37) (a). Atomic number=24. Electronic configuration 2, 8, 8, 6 and shells are K, L, M, N. "n" for last shell number M=**4** and outer most shell electron number **6**. So element will be placed at 4th period,6th Group. The element is Cr.

Or $1s^2 2s^2 sp^6 3s^2 3p^6 4s^2 4p^4$
Here sub shell number 4 and the outer most electron is (2+4=6). So element will be placed at 4th period, 6th Group. The element is Cr.

38) (c) Na and K are in the same group .K is in 4rd group compare to a 3rd group of Na.

K has a larger size of an atom. It will lose electron more easily from the outer most shell. Mg and Ca are smaller in size. Increasing order of atomic size these elements are Mg, Na, Ca, K.

39) **(a)** Electron affinity is the energy released when an electron is added to a neutral atom to from a negative ion.

40) **(a)** Metallic character increase down the group.

41) **(c)** Lanthanides are started from cerium (Ce) of atomic number 58.

42) **(b)** Ionization potential increases due to increase in nuclear charge along period from left to right.

43) **(d)** Group 2-4 are metal. Group 11 & 12 also contain metal and group 13 having metalloid Boron.14,15 and 16 group having metalloids.

44) **(b)** Mendeleev periodic table contain 6 periods and 8 groups.

45) **(a)** Ionization potential increases due to increase in nuclear charge along period from left to right.

46) **(c)** Al valence 3, Oxygen valence 2. So eka-Aluminum oxide formula will E_2O_3.

   E   O
   3   2
   Formula E_2O_3.

47) **(c)** They are called inner transition elements.

48) **(b)** Z=3 is Li, Z=7 is N_2, Z=12 is Mg and Z=13 is Al. Nitrogen is nonmetal and forms an acidic oxide.

49) **(a)** Lithium located in the second period and first group. Its alkali metal group as all metal oxide of this group makes alkali when mixed with water.

50) **(c)** Due to larger in size, it is less attractive to the extra electron.

51) **(b) Ionization energy** is qualitatively defined as the amount of **energy** required to remove the most loosely bound electron, the valence electron, of an isolated gaseous atom to form a cation. From top to bottom of a group atomic size increases so it helps to remove loosely bound energy.

52) **(d)** Scandium.

53) **(c)** All lanthanides outermost sub shell is f-sub shell.

54) **(a)** All inner transition elements (lanthanides and actinides) are metal.

55) **(c)** Hg is liquid metal. All metal oxides are alkaline in nature. Gallium is a metal which melt at human body temperature.

56) **(c)** Astatine in only natural radioactive Nobel gas element.

57) **(b)**.

58) **(d)** Radioactivity is not a periodic property.

59) **(b)** Modern periodic table has 4 blocks. s, p, d & f blocks.

60) **(d)** Na and K are in same group. K is in 4th group compare to 3rd group of Na.
K has larger size of atom. It will lose electron more easily from outer most shell. Mg and Ca are smaller in size. Increasing order of atomic size these elements are Mg, Na, Ca, K.

61) **(c)** Iodine is a solid halogen. F and Cl are gaseous halogen. Bromine is liquid halogen.

62) **(d)** Mosely proposed that elements property changes according to increase in atomic number.

63) (b). Number of isotopes and radioactivity are not periodic property.

64) (b). Lithium located at first group of 2nd period.

65) (d). Francium located at 7th period, 1st group. Periodic table has maximum 7th period.

66) (c). 7th period till date not filled.

67) (d). Along a period from left to right atomic radius decreases due to increase in nuclear charge, Metallic property decreases and valence changes.

68) (d). from top to bottom number of shell increases so atomic radius increases.

69) (a). B, Si, Ge, As, Te and Po are the metalloids in periodic table.

70) (a). Electro negativity is a measure of the tendency of an atom to attract a

bonding pair of electrons. In the same period nuclear charge increases so tendency to attract a bonding pair electron increases.

71) (a). Last elements of each group are Nobel gas and its last shell is filled completely so valence is zero.

72) (d). Nobel gases were invented after Mendeleev. Mendeleev subdivided each group but in modern periodic table, it has no subdivision. Isotopes are not invented during Mendeleev. Hydrogen was located in first group in Mendeleev periodic table.

73) (d). Outermost shell electron indicate group position so valence remain same for all element in a group.

74) (b). Generally, elements which contains 4 electrons in outermost shell having catenation property. Examples are C, Si, Ge, Sn.

75) (a). Heavy elements not preciously follow Octave rule.

76) (b). Elements with outermost shell electron contain 7 located at 17th group which is halogen group.

77) (d). Be (Beryllium) is Metal, Arsenic (As) is metalloid & Phosphorus (P) is non-metal.

78) (c). Atomic number 2+8+8=18. Its Argon.

79) (b).
A: Atomic Number 3
Electronic configuration 2, 1.
B: Atomic Number 12
Electronic configuration 2, 8, **2**.
C: Atomic Number 20
Electronic configuration 2,8,8,**2**.
D: Atomic Number 22
Electronic configuration 2,8,8,4.

So Element **B** & **C** having same Group (2) but periods are 3rd & 4th respectively.

80) (a) Outer most shell is completely filled with electron so valence is zero.

Life Processes

1) **During defficiency of oxygen in tissues of human beings, Pyruvic acids converted into**

   a) Lactic acid.
   b) Acetic acid.
   c) Citric acid.
   d) Ethyl alcohol.

2) **Conversion of Pyruvic acid to lactic acid takes place due to lack of oxygen in**

   a) Mitochondria.
   b) Golgi body.
   c) Cytoplasm.
   d) Centriole.

3) **Food takes time to reach stomach after swallowing food**

   a) 5 sec.
   b) 6 sec.
   c) 15 sec.
   d) 30 sec.

4) **Urine normally contain**

   a) 10% urea.
   b) 5% urea.
   c) 3% urea.
   d) 2% urea.

5) **The inner lining of the stomach is protected by one of the following from hydrochloric acid**

   a) Pepsin.
   b) Mucus.
   c) Salivery amylase
   d) Bile.

6) **Which part of alimentary canal receives bile from the liver?**

   a) Duodenum.
   b) Ilium.
   c) Jejunum.
   d) Stomach.

7) **Energy stored in glucose is**

   a) 686 Cal/gm.
   b) 386 Cal/gm
   c) 1056 Cal/gm
   d) 186 Cal/gm.

8) **Which of the following statement(s) is wrong regarding autotrophs?**

   1) They produce carbohydrates from CO_2 and H_2O in the presence of sunlight.
   2) They produce the same molecule of O_2 as same as CO_2 has absorbed.
   3) They stored carbohydrates in the form of starch.
   4) Starch color becomes sky when treated with iodine.

   a) Only 2 is wrong.
   b) 1 & 2 are wrong.
   c) 3 & 4 are wrong.
   d) Only 4 is wrong.

9) **Cresco graph is used to measure**

a) The growth of plant.
b) The growth of bacteria.
c) The growth of animal.
d) Respiration rate of the plant body.

10) The filtration unit of the kidney is

a) Nephrons.
b) Ureter.
c) Henley loop
d) Neurons.

11) The exit of food from the stomach is regulated by

a) Sphincter muscle.
b) Gastrocnemius muscle.
c) Deltoid Muscle.
d) Ciliary muscle.

12) Which of the following is not contains in Glomerulus

1. Bowman's capsule.
2. Renal artery.
3. Collecting duct.
4. Henley loop.

a) 1 & 3.
b) 1 & 2.
c) 2 & 3.
d) All.

13) Tricuspid valve open towards

a) Right atrium to the right ventricle.
b) Right Ventricle to pulmonary arteries.
c) Left atrium to left ventricle.
d) Left Ventricle to pulmonary veins.

14) Mitral valve (Bicuspid) open towards

a) Right atrium to the right ventricle.
b) Right Ventricle to pulmonary arteries.
c) Left atrium to left ventricle.
d) Left Ventricle to pulmonary veins.

15) Pulmonary valve open towards

a) Right atrium to the right ventricle.
b) Right Ventricle to pulmonary arteries.
c) Left atrium to left ventricle.
d) Left Ventricle to pulmonary veins.

16) Tricuspid valve open towards

a) Right atrium to the right ventricle.
b) Right Ventricle to pulmonary arteries.
c) Left atrium to left ventricle.
d) Left Ventricle to pulmonary veins.

17) Animals like human beings types of nutrition are

a) Saprophytic
b) Autotrophic.
c) Parasitic.
d) Holozoic

18) Which chamber of heart supply oxygenated blood to the body

a) Right Ventricle.
b) Left auricle.
c) Right auricle.
d) Left Ventricle.

19) Which chamber of heart supply carbonated blood to the lungs

a) Right Ventricle.
b) Left auricle.
c) Right auricle.
d) Left Ventricle.

20) Sub lingual gland located at

a) Below the tongue.
b) Beside the mandible.
c) Behind buccal cavity.
d) Stomach.

21) Which of the following node initiate heart beat?

a) AV node.
b) SA Node.
c) Purkinje fiber.
d) Bundle of His.

22) Our kidney is having

a) 100% stand by.
b) 200% stand by.
c) 50% stand by.
d) No stand by.

23) Complete digestion of protein, carbohydrate and fat takes place at

a) Stomach.
b) Large intestine.
c) Small intestine.
d) Rectum.

24) The main function of the large intestine is

a) To absorb water from the unabsorbed food.
b) Absorb fat related food.
c) Absorb unabsorbed food in the small intestine.
d) Convert unabsorbed food to absorbable by adding water.

25) Largest gland of the human body is

a) Liver.
b) Kidney.
c) Pancreas.
d) Gall bladder.

26) Oxygenated blood from lungs comes to the left atrium through

a) Vena cave.
b) Pulmonary arteries.
c) Capillary.
d) Pulmonary veins.

27) The right part of human heart contains.

a) Deoxygenated blood.
b) Oxygenated blood.
c) Mixed blood.
d) Lymph.

28) Following enzyme helps to clot blood

a) Fibrin.
b) Thrombokinase.
c) Bilirubin.
d) None of the above.

29) Aortic valve present in

a) Ventricle to the atrium.
b) Atrium to Ventricle.
c) Ventricle to the aorta.
d) Ventricle to the Pulmonary artery.

30) Pepsin function after the presence of

a) Hydrochloric acid.
b) Bile.
c) Trypsin.
d) Lipase.

31) The outermost layer of the heart is

a) Pleura.
b) Pericardium.
c) Ergosterol.
d) Meninges.

32) Theory of vitalystic was proposed by

a) J C Bose.
b) S C Bose.
c) Homi jahangir Bhaba.
d) Mendel.

33) Which of the following enzyme first mix with the food in the digestive tract.

a) Amylase.
b) Pepsin.
c) Trypsin.
d) Maltose.

34) Villi present in

a) Stomach.
b) Large intestine only.
c) Small intestine only.
d) Both small & large intestine.

35) Energy stored in the food as

a) ADP.
b) ATP.
c) DDT.
d) Glucose.

36) The muscle of Ventricle compare to the atrium

a) Thicker.
b) Less thick.
c) Are same in thickness.
d) Left and right side of the heart are different in thickness.

37) The outermost layer of lungs is called

a) Pleura.
b) Pericardium.
c) Ergosterol.
d) Meninges.

38) Tissue fluid is

a) Blood.
b) Lymph.
c) Water.
d) Both (a) and (b)

39) Watering of mouth is actually

a) Secretion of mucus.
b) Secretion of water.
c) Secretion of enzymes.
d) Secretion of the hormone.

40) Lub and dub is related to

a) Ear.
b) Heart.
c) Stomach.
d) Kidney.

41) Which elements are necessary for plant body for proper growth?

a) Nitrogen
b) Phosphorus.
c) Potassium.
d) All of the above.

42) A normal adult human being systolic and diastolic pressure is

a) 70mmHg & 150mmHg respectively.
b) 80mmHg & 120mmHg respectively.
c) 70mmHg & 100mmHg respectively.
d) 90mmHg & 150mmHg respectively.

43) The blood pigment of Mollusks animals is

a) Hemoglobin.
b) Hemerythrin.
c) Hemocyanin.
d) No pigment is present.

44) Plant respiration takes place

a) During photo synthesis.
b) Throughout the day.
c) During night only.
d) Depends on the type of plant.

45) During respiration glucose first converted to

a) Pyruvic acid.
b) Acetic acid.
c) Lactic acid.
d) Ethyl Alcohol

46) Lub sound occur when

a) Systole occurs, mitral & Tricuspid Open.
b) Diastole occur, mitral & Tricuspid close.
c) Systole occur, mitral & Tricuspid close.
d) Diastole occur, mitral & Tricuspid open.

47) Which of the following have two chambered heart

a) Frog.
b) Cow.
c) Crocodile
d) Fish.

48) Normal adult human being blood sugar range is

a) 90 mg to 120mg per 100ml blood.

b) 70 mg to 120mg per 100ml blood.
c) 100 mg to 120mg per 100ml blood.
d) 150 mg to 180mg per 100ml blood.

49) The function of mucus in the digestive system is

a) To digest carbohydrates.
b) To digest fat.
c) To protect the stomach from HCl.
d) To kill bacteria only.

50) Most evaluated animal having the following number of heart chambers

a) 2
b) 3
c) 4
d) 5

51) _____ works as a filter in lungs

a) Alveoli.
b) Bronchioles.
c) Bronchi.
d) Trachea.

52) Valves present in

a) Arteries.
b) Veins.
c) Capillaries.
d) None of the above.

53) Photosynthesis takes place

1. During daytime.
2. During night.
3. Throughout the day.
4. According to available light intensity.

a) 1 & 2.
b) 3 only.
c) 1 & 4.
d) 3 only.

54) Which of the following organ is bean shaped?

a) Kidney.
b) Liver.
c) Pancreas.
d) Gall bladder.

55) The optimum temperature of photosynthesis is?

a) 20 Deg C
b) 30 Deg C
c) 25 Deg C
d) 35 Deg C

56) Frogs having the following number of heart chambers.

a) 2
b) 3
c) 4
d) 5

57) Which one carries oxygenated blood?

a) Pulmonary arteries.
b) Pulmonary vein.
c) Capillaries.
d) Both 2 and 3.

58) Draining of excess fluids from intercellular space is carried out by

a) Blood.
b) Lymph.
c) Both blood and lymph.
d) The cell membrane of adjacent cells.

59) Respiration of our body is taking place

a) During daytime.
b) During night.
c) During high-energy requirement.
d) Throughout the day.

60) Which one is the building blocks of protein

a) Amino acids.
b) Fatty acids.
c) Glucose.
d) Starch.

61) Which one is the building blocks of carbohydrate

a) Amino acids.
b) Fatty acids.
c) Starch.
d) Glucose.

62) Which one is the building blocks of fat

a) Amino acids.
b) Glucose.
c) Fatty acids.
d) Starch.

63) Glucose stored in the form of ____ in our body

a) Starch.
b) Glycogen.
c) Sucrose.
d) Lactose.

64) What is the color of starch in iodine solution?

a) Green.
b) Red.
c) Yellow.
d) Blue.

65) Pores of leaves open and close carried out by

a) Lenticels.
b) Guard cells.
c) Xylem.
d) Phloem.

66) Which food is called protein saving food

a) Glucose.
b) Carbohydrates.
c) Fatty acids.
d) Amino acids.

67) Due to excessive starvation following compound found in urine

a) Bilirubin.
b) Biliverdin.
c) Ketone body.
d) Albumin.

68) The main execratory product of aves is

a) Ammonia.
b) Urea.
c) Uric acid.
d) Ketone body

69) Pacemaker used when _____ node fail to function properly

a) Sino auricular node.
b) Auricle ventricular node.
c) Both the above.
d) None of the above.

70) Lipase enzyme works on

a) Protein.
b) Carbohydrate.
c) Fat.
d) Vitamins.

71) Life of RBC in our body is

a) 100 days.
b) 7 days.
c) 30-40 days.
d) 120 days.

72) Which of the following is not related to the kidney?

1. Filtering.
2. Transferring.
3. Storage.
4. Renin.

a) 3 & 4.
b) 2 & 3.
c) 4 only.
d) None

73) Which of the following juices is secreted by the pancreas

1. Renin.

2. Pepsin.
3. Bile juice.
4. Lipase.
5. Trypsin

a) 1, 3 & 4.
b) 2, 4 & 5.
c) 2 & 3.
d) 2, 3 & 4.

74) Following blood cell helps to clot blood

a) Platelets.
b) Basophils.
c) Eosinophils
d) RBC.

75) Artificial Dialysis method used due to the failure of the following organ

a) Lungs
b) Heart
c) Kidney.
d) Diaphragm.

76) Protein contained in lymph compared to blood is

a) More.
b) Less.
c) Slightly more.
d) Some place its more and some place its less.

77) Bile juice is produced by

a) Gal bladder.
b) Liver.
c) Pancreas.
d) Stomach.

78) What is the role of bile in the digestive system

a) It emulsified the fat.
b) Assimilated the fat.
c) It helps to absorb fat.
d) It helps to convert low-density fat to high density fat.

79) Following vitamin helps to clot blood

a) A
b) D
c) E
d) K.

80) Which element helps to clot blood

a) Na
b) Ca
c) Zn
d) Fe

81) Which organ supported by pacemaker?

a) Lungs.
b) Kidney.
c) Heart.
d) Ear.

82) Bile juice is secreted by

a) Gal bladder.
b) Liver.
c) Pancreas.
d) Stomach.

[A COMPLETE GUIDE TO MCQ]

CHAPTER: 06

Answers & Explanations

1) **(a).** Lactic acid creates muscle fatigue.

2) **(c).** it takes place in anaerobic respiration.

3) **(b).** 6 sec require transferring food from buccal cavity to stomach.

4) **(d).** 95% Water + 3% salts +2% urea = urine.

5) **(b).** Mucus protects stomach layer from HCl corrosive action.

6) **(a).**

7) **(a).**

8) **(d).** Starch color become blue in presence of iodine and all other statements are correct.

9) **(a).** The growth of plant measured by Cresco graph and it was in vented by J C Bose.

10) **(a).** The nephron is the building unit of the kidney.

11) **(a).** Sphincter muscle by peristaltic movement transfer food.

12) **(c).** Collecting duct and Henley loop is outside the glomerulus.

13) **(a).**

14) **(c).** It allows blood to pass from the upper left atrium side to the lower left ventricle side.

15) **(b).** The opening of pulmonary valve sends deoxygenated blood from right ventricle to lungs to purify.

16) **(a).** The opening of Tricuspid valve sends deoxygenated blood from right atrium to right ventricle.

17) **(d).** Holozoic means the animals take whole food and after chewing, it goes for digestion.

18) **(d).** Left ventricle supply oxygenated blood to the body through mitral valve during the systolic stroke of heart.

19) **(a).** The right side of heart contains deoxygenated blood.

20) **(a).** Its mainly produces mucus.

21) **(b).** SA node is the creator of electric impulse.

22) **(a).** One kidney of our body is sufficient for extractor system but they work on 50% load sharing.

23) **(c).**

24) **(a).**

25) **(a).** The liver is the largest gland of the human body and could regenerate itself.

26) **(d).** Pulmonary veins only carry oxygenated blood to the left atrium.

27) **(a).** Right part of heart collect deoxygenated blood from the whole body and then send it to lungs for oxygenating the same.

28) **(b).**

29) **(c).** The aortic valve is located between the aorta and the heart's left ventricle. The pulmonary vein delivers oxygenated blood to the heart's left atrium & then it passes through the mitral valve and enters into the left ventricle.

30) **(a). Pepsin** works in the acidic medium.

31) **(b). Pleura** is the outermost layer of lungs. The pericardium is outer most layer of the heart. Ergosterol is a layer on the skin where vit. D synthesized. Meninges is the outermost layer of the brain.

32) **(a).**

33) **(a).** Salivary Amylase secrets inside the buccal cavity and it first mixes with food.

34) **(c).** The villi and the microvilli increase intestinal absorptive surface area approximately 30-fold and 600-fold, respectively, providing exceptionally efficient absorption of nutrients in the lumen.

35) **(b).** ADP (Adenosine Di phosphate converted to Adenosine tri phosphate by addition of pi)

36) **(a).** Ventricle muscles are thicker due to high systolic pressure it has to sustain.

37) **(a).** Pleura is the outermost layer of lungs. The pericardium is outer most layer of the heart. Ergosterol is a layer on the skin where vit. D synthesized.

38) **(b).** LYMPH is called tissue fluid as it produced at the space between the cells. It's one type of circulation system of our body.

39) **(c).** Please note abnormal watering occurs during rabies and pellagra dieses.

40) **(b)** These are turbulence sound produced during systole and diastole of the heart chambers.

41) **(d)** NPK are the elements for the growth of plant body.

42) **(b).** Normal systolic and diastolic pressure range if 80mmHg to 120mmHg.

43) **(c).** Hemoglobin: Higher order animal, Fe pigment. Hemerythrin: For insects, Iron based pigment. Hemocyanin: Copper based pigment.

44) **(b).** Respiration required to carry out throughout the day to supply energy.

45) **(a).** Pyruvate is the output of the anaerobic metabolism of glucose known as glycolysis.

46) **(c).** It is caused by turbulence caused by the closure of mitral and tricuspid valves at the start of systole.

47) **(d).** Fish: Two chambered. Cow and Crocodile: 4 Chambered, Frog: Three chambered according to evolution.

48) **(b).** The normal range is 70 mg to 120mg per 100ml of blood. More than the specified limit is called hyperglycemia and less than limiting value is called hypoglycemia.

49) **(c).** Mucus helps to protect the stomach from corrosive effects of HCl.

50) **(c).** All top-level animals' number of the heart chamber is 4.

51) **(a).** Alveoli are the building unit of lungs.

52) **(b).** Veins have a valve but arteries not having a valve.

53) **(c).** Photosynthesis takes place during daytime and with the presence of correct intensity of light at any time throughout the day.

54) **(a).**

55) **(c).** Hg is liquid metal. All metal oxides are alkaline in nature. Gallium is a metal which melts at human body temperature.

56) **(b).** Amphibians number of chamber of hearts is 3.

57) **(b).** Generally, vein carries deoxygenated blood but it opposite to pulmonary vein.

58) **(b).** Lymph is another circulatory system of our body which drain excess fluids from intercellular space.

59) (d). Respiration requires throughout the day to supply energy for function body.

60) (a).

61) (d). Glucose is the smallest unit of carbohydrates.

62) (c).

63) (b). By glycogenesis method glucose converted to Glycogen and stored in muscle & liver.

64) (d).

65) (b). Guard cell helps to open & close pores. Xylem and Phloem transfer water and food respectively. Lenticels are pores of stem cover.

66) (b). Taking only carbohydrate food could maintain our body health, so it called protein saving food.

67) (c). Excessive starvation breaks protein and it produces ketone body which filtered out by the kidney.

68) (c).

69) (a). SA node sends out regular electrical impulses from the top chamber (the atrium) causing it to contract and pump blood into the bottom chamber (the ventricle).

70) (c). Lipase enzyme produced by the pancreas and act on fat.

71) (d). The life span of RBC is 120 days.

72) (d). All are the function of Kidney and Renin enzyme produced by the juxtaglomerular cell in the kidney.

73) (b). Renin: Juxtaglomerular cell of kidney, pepsin, lipase & Trypsin: Pancreas, Bile juice: Produced by the liver and secretes by Gall bladder.

74) (a) Platelets produce thrombokinase that helps to clot blood by creating fibrin.

75) (c). Artificial dialysis is carried out when both kidneys is not functioning correctly.

76) (b). The fluid leaked from the cell is converted as lymph so its protein content is less.

77) (a). Bile juice is produced by Gall bladder and secrets by liver.

78) (a). Bile helps to emulsify the fat and helps to digest.

79) (d). Vitamin K helps to clot blood.

80) (b). Ca^{++} ion helps to clot.

81) (c). Heart uses pacemaker when SA node is not function properly.

82) (b). Bile juice secretes by liver but produces by gall bladder.

Control & Coordination

1) The building unit of nervous system is

a) Neuroglia.
b) Neuron.
c) Receptor.
d) Dendron.

2) ____ hormone is responsible for dark color of areola

a) Auxins.
b) Glucagon.
c) Melatonin.
d) Androgen.

3) Receptors are usually located at

a) Sense organs.
b) Mussels.
c) Skin.
d) brain.

4) Which of the following is dual gland

a) Pancreas.
b) Pituitary.
c) Pineal body.
d) Liver.

5) Who coined the term "Hormone"

a) Watson and Crick.
b) Horgobind Khurana
c) Hook.
d) Bayliss & Starling.

6) Which gland is called master gland

a) Pituitary.
b) Pineal body.
c) Liver.
d) Pancreas.

7) Which system maintains homeostasis (equilibrium state of systems)?

a) Endocrine ystem.
b) Exsocrine system.
c) Nervous system.
d) All of the above.

8) Which hormone triggers the fall of mature fruits and leaves

a) Cytokinins.
b) Abscissic acid.
c) Ethylene.
d) Gibberellins.

9) C.S. Fluid is found in

a) Brain.
b) Spinal cord.
c) Both (a) and (b).
d) At the junction of two nerve.

10) The father of reflex action theory is

a) Pavlov.
b) Mendel.
c) J C Bose.
d) Sleiden.

11) **Spinal cord is a part of following nervous system.**

a) Peripheral nervous system.
b) Central nervous system.
c) Automatic nervous system.
d) Voluntary nervous system.

12) **Left and right part of the brain is connected by**

a) Pons.
b) Medulla oblongata.
c) Corpus callosum.
d) Corpus luteum

13) **Which hormone helps to increase heart beat during emergency**

a) Cardiac hormone.
b) SA node.
c) Adrenalin.
d) Progesterone.

14) **Reflex of eyes and ear control by**

a) Pons.
b) Hind brain.
c) Mid brain.
d) Fore brain.

15) **Auxins stored in**

a) Shady side of shoot.
b) Opposite of shady side of shoot.
c) In between light and shaded part of shoot.
d) At the bottom most part of root.

16) **Hypothalamus is the part of brain which interlinked between**

a) Nervous and endocrine system.
b) Endocrine and exocrine system.
c) Mid and fore brain.
d) Control voluntary movements.

17) **Hormones are grouped into ____ class.**

a) 2.
b) 3.
c) 4.
d) Only one.

18) **Sex hormones are one types of____ hormone.**

a) Steroid
b) Peptide.
c) Amine
d) none of the above.

19) **STH is one type of**

a) Local hormone.
b) Tropic hormone.
c) Steroid hormone.
d) None of the above.

20) **Female reproductive organ growth related with**

a) Progesterone hormone.
b) Testosterone hormone.
c) Oestrogen hormone.
d) Luteinizing hormone.

21) **Pineal body located at**

a) Below Pancreas.
b) Above kidney.
c) Brain.
d) Near to spleen.

22) **Which hormone helps to control water content in urine**

a) ADH
b) TSH
c) STH
d) Adrenalin

23) **The chemical responsible for transferring electric pulse at synapse is**

a) Acetylcholine.
b) Acetic acid.
c) Ethylene.
d) Thyroxin.

24) Which hormone is essential for cell division in plant body

a) Ethylene.
b) Cytokinin.
c) Acetylcholine.
d) Auxins.

25) Thalamus is a part of

a) Fore brain.
b) Hind brain.
c) Spinal cord.
d) Mid brain.

26) Guard cell function controlled by

a) Hormone.
b) Pheromone.
c) Minerals.
d) Water.

27) During child birth following hormone is essential

a) Estrogen.
b) Testosterone.
c) Oxytocin.
d) STH.

28) Which hormone also called personality hormone

a) Thyroxin.
b) Adrenaline.
c) Noradrenalin.
d) androgen

29) Almost all human body cell could be targeted by

a) Adrenalin hormone.
b) Thyroid hormones.
c) Sex hormones.
d) Growth hormones.

30) Adrenal gland is located

a) On liver.
b) On kidney.
c) On larynx.
d) On pancreas.

31) For milking baby _____ hormone take active part

a) Luteinizing
b) Ganado tropic.
c) Thyroid.
d) Oxytocin.

32) Following hormone, regulate glucose content in blood.

a) Insulin & Glucagon
b) T4 and T3
c) Adrenalin and GTH.
d) Progesterone & Estrogen.

33) For Bonsai following hormone is required

a) Auxins
b) Gibberellin.
c) Cytokinin.
d) Abscisic acid.

34) Cytokinin is found in

a) Seeds.
b) Flower.
c) Leaves.
d) Stem.

35) The advantages of hormonal communications over electrical communications are

1. Its reach all the cells where nerve cells are not connected.
2. Electrical communications are slower than chemical communication.
3. Nerve cells cannot generate & transmit impulse continuously.

a) All are correct.
b) 2 & 3 are correct.
c) 1 & 3 are correct.
d) 1 & 2 are correct.

36) Regulation of calcium content in bone is carried out by following hormone

a) Thyroid.
b) Calcitonin.
c) STH
d) GTH

37) Insulin hormone is used to treat

a) Diabetic mellitus.
b) Diabetic insipidus.
c) Both (a) and (b).
d) Goiter.

38) Growth of stem is controlled by

a) Auxins.
b) Gibberellin.
c) Cytosine.
d) Ethylene.

39) From islets of Langerhans cells off pancreas following twin hormones are secrets

a) T4 and T3
b) Adrenalin and GTH.
c) Insulin & Glucagon
d) Progesterone & Estrogen.

40) Respiration is controlled by

a) Medulla.
b) Pons.
c) Cerebrum.
d) Thalamus.

41) Spinal cord is surrounded by

a) Pleura.
b) Meninges.
c) Pericardium.
d) CSF.

42) Following hormone helps for metamorphosis

a) Thyroid.
b) Thyroxin.
c) Adrenalin.
d) STH.

43) Goiter is deficiency of

a) Hormone.
b) Minerals.
c) Food.
d) Water.

44) Which hormone contains Iodine?

a) Adrenalin.
b) Thyroxine.
c) STH
d) Adrenalin.

45) Honey bee could recognize each other due to

a) Hormone.
b) Developed peripheral nervous system.
c) Pheromone.
d) Acetylcholine.

46) The neurons of white matter of our brain is

a) Non-Myelineted type.
b) Myelineted type.
c) Both (a) & (b) type.
d) Ranvier node free nerve.

47) Ranvier node found in

a) Myelineted type.
b) Non-Myelineted type.
c) Both (a) & (b) type.
d) Neuroglia.

48) Following gland is responsible for blood sugar control

a) Pancreas.
b) Liver.
c) Adrenal.
d) Thyroid.

49) Long term chemical control carried out by

a) Hormones.
b) Nerve.
c) Both (a) and (b).
d) Synapse.

50) _____ hormone is called emergency hormone

a) Testosterone.
b) Luteinizing.
c) Progesterone.
d) Adrenalin.

51) Parthenocarpy is a method of producing

a) Fruits without flower.
b) Fruits without seeds.
c) Fruits of bigger size.
d) New plant without seed.

52) Which hormone require Iodine

a) Adrenalin.
b) Thyroxin.
c) STH
d) Adrenalin.

53) Human nervous system is divided into

a) 2
b) 3
c) 4
d) None of the above.

54) Growth of pollen tube towards the ovule is called

a) Geotropism.
b) Chemotropism.
c) Phototropism.
d) Thermo tropism.

55) Nissle's granules found in

a) Liver.
b) Nerve cells.
c) Eye lenses.
d) Ear drum.

56) Due to absence of following hormone hyperglycemia could happen

a) Insulin.
b) Glucagon.
c) Thyroid.
d) Adrenalin.

57) Reflex center of the brain is

a) Thalamus.
b) Hind brain.
c) Pons.
d) Medulla oblongata.

58) Schwann cells are found

a) Above myelin sheath.
b) Below myelin sheath.
c) Above Meninges.
d) Inside spinal cord.

59) Dwarfism is the effect of

a) Less production of STH.
b) Less production of GTH.
c) More production of STH.
d) More production of GTH.

60) Following movement found in roots

a) Geotropism.
b) Hydrotropism.
c) Chemotropism.
d) Phototropism.

61) Adrenalin gland located at

a) Pancreas.
b) Kidney.
c) Larynx.
d) Liver.

62) Gustatory receptors detects

a) Taste.
b) Smell.
c) Image.
d) Sound waves.

63) Due to absence of following hormone hypoglycemia could happen

a) Insulin.
b) Thyroid.
c) Glucagon.
d) Adrenalin.

64) Which element is essential for thyroxin synthesis

a) Iodine.
b) Iron.
c) Copper.
d) Calcium.

65) Olfactory receptors detects

a) Taste.
b) Smell.
c) Image.
d) Sound waves.

66) Following part of brain is associated with endocrine and nervous system both

a) Hypothalamus.
b) Pons.
c) Thalamus.
d) Cerebrum.

67) Which of the following is/are not a part of reflex action?

1) Receptor.
2) Effectors.
3) Spinal cord.
4) Neuroglia.
5) Motor nerve.
6) Synapse.

a) 6 only.
b) 3 & 6.
c) 1,2 & 5.
d) 4 & 6.

68) Which hormone helps to absorb and control water in urine

a) ADH
b) ACTH
c) FSH
d) TSH

69) The coating of brain is called

a) Pericardium.
b) Meninges.
c) Pleura.
d) None of the above.

70) Followings is/are phyto hormone(s)

1) Auxins.
2) Ethylene.
3) Cytokinin.
4) Florigen.

a) 1 only
b) 1 & 3 only
c) 1 ,3 & 4 only
d) All of the above.

71) Which gland is located lowermost part of the body

a) Testes.
b) Adrenalin.
c) Thyroid.
d) Pineal body.

72) Sudden increase in body weight up to 5 kg is the malfunction of following hormone

a) Melatonin.
b) Thyroxin.
c) Adrenalin.
d) Oestrogen.

73) Typing without seeing keys is an example of

a) Unconditional reflex action.
b) Unconditional reflex action.
c) Semi conditional reflex action.
d) It's not any type of reflex action.

74) Which of the following is not a tropic hormone

a) LH
b) FSH
c) STH
d) ACTH

75) Which of the followings are controlled by peripheral nervous system

1) Vomiting.
2) Twitching of eyelids.
3) Snoring.
4) Chewing.

a) 1 & 2 only.
b) 2 & 3 only.
c) 3 only.
d) All of the above.

76) The boney box which protect brain is called

a) Meninges.
b) Medulla.
c) Cranium.
d) Corpus lutetium.

77) Chemical signal transmitted at synapse from

a) Axon of one neuron to cyton of other neuron.
b) Axon of one neuron to Dendron of other neuron.
c) Dendron of one neuron to cyton of other neuron.
d) Axon of one neuron to axon of other neuron.

78) Spinal cord is originates from

a) Cerebellum.
b) Cerebrum.
c) Medulla.
d) Pons.

79) The largest part of the brain is

a) Cerebellum.
b) Cerebrum.
c) Pons.
d) Thalamus.

80) Which hormone helps to decide, "Fight or Flight"?

a) Adrenalin.
b) Melatonin.
c) Progesterone.
d) Testosterone.

[Answers & Explanations]

1) **(b).** Neuroglia is the non-neuronal cells in CNS & PNS. Neuron is the building unit of nervous system. Receptor receives signal and Dendron is a part of neuron.

2) **(c).** Auxins is plant hormone. Glucagon maintain sugar level and Androgen is steroidal hormone that regulate the development and maintenance of male characteristics. Melatonin secrets from pineal body of brain.

3) **(a).** Receptors are sense from sense organ.

4) **(a).** pancreas is called dual gland as it secrets enzyme (amylase, lipase) and hormone (insulin & glucagon) both. Others are secrets either enzyme or hormone.

5) **(d).** In 1902 and named first hormone as secretin.

6) **(a).**

7) **(a).**

8) **(b).**

9) **(b).** C.S fluid is **Cerebral spinal fluid**.

10) **(a).** Ivan Pavlov is the father of reflex action.

11) **(b).** Spinal cord & brain altogether form Central Nervous System. Nerves and ganglia outside the brain and spinal cord is form peripheral nervous system. Option c & d are part of PNS.

12) **(c).** Corpus callosum connect right & left part of brain. Corpus luteum is the part of ovary which secrets hormone. Pons connect upper & lower part of the brain. Medulla oblongata control function of heart and lungs.

13) **(c).** Adrenalin is also called emergency hormone. SA node is not a hormone and there is no hormone named cardiac hormone. Progesterone hormone is female hormone, which develop female reproductive organs.

14) **(c).**

15) **(a).** Auxins stored when light is less so that cell of that part could develop more quickly to control phototropism.

16) **(a).**

17) **(c).** Peptide (Ex Glucagon), Steroid (testosterone) and Amino (thyroxin) hormone.

18) **(a).**

19) **(b).** Tropic hormones are secrets from one gland and act on other organ. STH secrets from pituitary but works on all body cell.

20) **(c).**

21) **(c).** Pineal body located behind the pituitary gland.

22) **(a).** Anti-diuretic hormone helps to absorb extra water from urine before it goes to bladder.

23) **(a).** It's called neurotransmitter.

24) **(b).** Ethylene helps to ripe fruit. Cytokinin is essential for plant body cell division.

25) **(a).**

26) **(d).** When water content increase Guard cell open the port and when water contain reduced its port closed.

27) **(c).** Oxytocin helps to contract wombs during delivery of babies' and helps to start lactation.

[A COMPLETE GUIDE TO MCQ] CHAPTER: 07

28) (a).

29) (b).

30) (b).

31) (d). **Luteinizing** hormone also known as lutropin which triggers ovulation & development of corpus lutetium. Oxytocin helps to trigger lactation.

32) (a). Insulin reduce glucose level in blood & increase the same by glucagon. T4 and T3 control metabolism of calcium and phosphorous.

33) (d). Abscisic acid helps to inhibits plant growth.

34) (a). Cytokinin is found in coconut and it's a seed also.

35) (a). **Nerve cells** transmits pulse intermittently with high frequency.

36) (b). Calcitonin lowers blood calcium levels by suppressing osteoclast activity in the bones and increasing the amount of calcium excreted in the urine.

37) (a) **Diabetic mellitus** is high content of glucose in blood so it's controlled by insulin hormone which secrets from pancreas. Insulin reduce glucose content in blood. Diabetic insipidus is control by ADH. It helps to absorb extra water filtered by glomerulus. Goiter is the effects of excess secretion of thyroid.

38) (b). Gibberellin hormone is helps to grow the stem.

39) (c). Islets of Langerhans are located inside our pancreas. They are composed of two types of cells, **alpha** cells and **beta** cells. Beta cells secrete insulin, which convert extra sugar present in our blood into glycogen and stores into liver. Alpha cells glucagon, which converts glycogen (stored sugar) into sugar when necessary.

40) (a).

41) (b). The meninges are three layers of protective tissue called the Dura mater, arachnoid mater, and pia mater.

42) (b). Thyroxin produce in thyroid and helps to metamorphosis and adrenalin is secrets from top of kidney, which helps to control body function during emergency. STH is helps to grow human body.

43) (a). Thyroxin hormone.

44) (b).

45) (c). Pheromone is a chemical produces and releases into the environment by animals (especially mammals) and insects which affecting behavior or psychology of others.

46) (b).

47) (d). Ranvier node is the gap between the Myelin sheath.

48) (a) See Serial No 37.

49) (a).

50) (d). Adrenalin hormone initiates quick reaction, which makes the individual to think & respond quickly to the stress.

51) (b). It is the method of food production without fertilization.

52) (b) Iodine is an constituents of thyroxin.

53) (a). **CNS** (Central nervous system) & **PNS** (peripheral nervous system)

54) (b) Growth or movement of a plant or plant part in response to a chemical stimulus is called chemotropism. Here pollen tube grows towards ovule.

55) (b).

[A COMPLETE GUIDE TO MCQ] CHAPTER: 07

56) (a). Hyperglycemia is the excess of sugar presence in blood and insulin convert excess glucose of blood to glycogen and then stored on liver.

57) (d).

58) (a).

59) (a). It is the hormonal and metabolic disorder, which cause dwarfism of skeletal and its control by STH. Its appear at during childhood.

60) (b). Roots grow towards water and its called hydrotropism.

61) (b). Adrenalin located above the kidney as shows like a hat on the kidney.

62) (a).

63) (c). Glucagon helps to increase glucose in blood and absence of glucagon will produce hypoglycemic condition.

64) (b).

65) (b).

66) (a).

67) (d). Neuroglia and **Synapse** is not the part of reflex action.

68) (a). See Serial no 37.

69) (b). See Serial no 41.

70) (d).

71) (a). Testes is located at the bottom most part of male body inside scrotum.

72) (b).

73) (b).

74) (a) LH is local hormone and all others are secrets from Pituitary and works at different location.

75) (a). Snoring and Chewing is not reflex action.

76) (c).

77) (b) Synapse is the junction of Dendron and axon.

78) (c).

79) (b).

80) (a). See Q-50.

How do Organisms Reproduce

1) **The basic events of reproduction is/are**

   1) Copying DNA.
   2) Mitosis.
   3) Meiosis.
   4) Mutation.

   a) 1 & 3
   b) 1, 3 & 4.
   c) 1 & 2.
   d) 1 only.

2) **Those animals which lay eggs are called**

   a) Vivi porous.
   b) Somniferous.
   c) Oviparous.
   d) Metamorphic.

3) **Zygote is**

   a) Haploid cell.
   b) Diploid cell.
   c) Either (a) or (b).
   d) None of the above.

4) **In diploid organisms meiosis occurs during**

   a) Gametes formation.
   b) Fetus development.
   c) Somatic cell division.
   d) Fertilization.

5) **Ova (human) contains**

   a) 22 somatic chromosome & one sex (X type) chromosome.
   b) 22 somatic chromosome & one sex (Y type) chromosome.
   c) 50% Ova contains 22 somatic chromosomes & 1 sex(X type) chromosome. Remaining 50 % contains 22 somatic chromosomes & 1 sex (Y type) chromosome.
   d) 22 sex chromosome & one somatic chromosome.

6) **Sperm (human)cell contains**

   a) 22 somatic chromosome & one sex (X type) chromosome.
   b) 22 somatic chromosome & one sex (Y type) chromosome.
   c) 50% sperm contains 22 somatic chromosomes & 1 sex(X type) chromosome. Remaining 50 % contains 22 somatic chromosomes & 1 sex (Y type) chromosome.
   d) 22 sex chromosome & one somatic chromosome.

7) **Those animals which give birth of young one is**

   a) Vivi porous.
   b) Somniferous.
   c) Oviparous.
   d) Metamorphic.

8) **Anther of flower contains**

[A COMPLETE GUIDE TO MCQ]

CHAPTER: 08

a) Pollen grains.
b) Carpel.
c) Ovules.
d) Sepals.

9) Number of chromosomes contain by human beings

a) 44 number.
b) 23 number.
c) 46 number.
d) 22 number.

10) Number of sex chromosomes contain by human beings

a) 44 number.
b) 2 number.
c) 46 number.
d) 22 pair.

11) Which of the following is the correct sequence of events of plants sexual reproduction?

a) Pollination, Fertilization, Seedling, embryo.
b) Seedling, Pollination, Fertilization, embryo.
c) Fertilization, Pollination, Seedling, embryo.
d) Pollination, embryo. Fertilization, Seedling,

12) The organism behind malaria reproduces by

a) Binary fission.
b) Multiple fission.
c) Budding.
d) Fragmentation.

13) Who pay role determining sex chromosome of foetus

a) Father.
b) Mother.
c) Both Father and Mother.
d) None, as it happen by natural selection.

14) Transmitted characters to the offspring's during reproduction shows

a) Only similarities with parents.
b) Only variations with parents.
c) Both similarities and variations of parents.
d) Neither similarities nor variation.

15) After grafting which portion will provide the stem

a) Scion.
b) Stock.
c) Either (a) or (b).
d) The best quality of (a) and (b)

16) Gametes of algae is transferred through _____ medium.

a) Water.
b) Tentacles.
c) Air.
d) Anther.

17) The organism behind Kala-azar reproduces by

a) Multiple fission.
b) Budding.
c) Binary fission.
d) Fragmentation.

18) After joining two oviducts of human female

a) Vagina formed.
b) Uterus formed.
c) Cervix formed.
d) Fallopian tube formed.

19) Asexual reproduction of spirogyra takes place by

a) Breaking up filaments into small buds.
b) Division of cell into two cells.
c) Division of cell into many cells.
d) Formation of new young one.

20) During grafting which one have better quality gene

How do Organisms Reproduce MCQ (class X) [Page 71]

a) Scion.
b) Stock.
c) Either (a) or (b).
d) Cannot possible to tell.

21) Scrotum located at

a) Bottom most par of male body.
b) Outside the male body.
c) Inside the male body.
d) Near to uterus.

22) Fragmentation occurs in

a) Single cell organisms.
b) Multi-cellular organisms.
c) Bothe (a) and (b).
d) Fetus only.

23) Which one is the correct sequence

a) Zygotes, Gametes, seedling, Embryo.
b) Zygotes, seedling, Gametes, Embryo.
c) Gametes, Zygotes, seedling, Embryo.
d) Embryo, Zygotes, Gametes, seedling.

24) Bryophyllum reproduces by

a) Buds.
b) Seeds.
c) Roots.
d) All of the above.

25) Which one is the correct sequence

a) Embryo, Zygotes, Gametes, Cleavages,
b) Cleavages, Zygotes, Embryo, Gametes.
c) Zygotes, Gametes, Cleavages, Embryo.
d) Gametes, Zygotes, Cleavages, Embryo.

26) In human body _____ gamete is motile.

a) Female.
b) Male.
c) Both male and female.
d) None of the above.

27) Following organism reproduce by regeneration

a) Hydra.
b) Amoeba.
c) Leishmania.
d) Plasmodium vivax.

28) Best mode of reproduction is

a) Budding.
b) Fragmentation.
c) Binary fission.
d) Reproduce by fertilizing gametes.

29) Fallopian tube is ligated in

a) Vasectomy
b) Tubectomy.
c) Shemale body.
d) None of the above.

30) Length of pollen tube depends on

a) Distance between pollen grain and upper surface of stigma.
b) Upper surface of sigma and lower surface of style.
c) Distance between pollen grain on upper surface of stigma & ovule.
d) Distance between pollen grain on anther and upper surface of stigma.

31) Which among the following is correct about unisexual flower?

1) They have stamen and pistil.
2) They have stamen or pistil.
3) They exhibit cross-pollination.
4) They cannot produce fruits.

a) All are correct.
b) 2 & 3 are correct.
c) 1 & 3 are correct.
d) 1 & 2 are correct.

32) Growing fetus derive nutrition from mother body by

a) Placenta.
b) Fallopian tube.
c) Uterus blood vessels.
d) Cervix.

33) Vas deferens is ligated in

a) Vasectomy
b) Tubectomy.
c) Shemale body.
d) None of the above.

34) Spores are unit of

a) Sexual reproduction.
b) Asexual reproduction.
c) Both (a) and (b).
d) None of the above.

35) Fruits are developed form of

a) Petals.
b) Ovule.
c) Ovary.
d) Stigma.

36) Which one is first step of production of foetus

a) Morella.
b) Blastula.
c) Gastrula.
d) Zygote.

37) Which of the following is an example of IUCD

a) Diaphragm.
b) Condom.
c) Copper-T.
d) Cervical cap.

38) Spore production is related to

a) Sexual reproduction.
b) Asexual reproduction.
c) Higher group of plants.
d) Not related to reproduction.

39) Rapidly offspring's are produced during

a) Sexual reproduction.
b) Asexual reproduction.
c) Artificial reproduction.
d) None of the above.

40) IUCD is used for

a) Contraception.
b) Improving fertility.
c) Protecting from STDs.
d) Miscarriage protection.

41) Male testes are located outside the body to

a) Maintain the growth of scrotum.
b) To develop sperm rapidly.
c) To control temperature lower than body temperature.
d) To protect it from urinal bladder.

42) Which of the following is not a STD

a) Syphilis.
b) Ghnoria.
c) Hernia.
d) AIDS

43) Vasectomy done in

a) Male body.
b) Female body.
c) Shemale body.
d) All of the above.

44) Lotus is pollinated by

a) Water.
b) Insects.
c) Birds.
d) Human.

45) Budding is a

a) Sexual reproduction method.
b) Asexual reproduction method.
c) Artificial reproduction method.
d) None of the above.

46) Function of prostate gland is

a) Provide fluid to move the sperm.
b) Nourishment of sperm.
c) Ejaculation of sperm.
d) All of the above.

[A COMPLETE GUIDE TO MCQ]

CHAPTER: 08

47) Each Ovary of flowers produces

a) Only one ovule.
b) Two ovules.
c) Three ovules.
d) One or two ovules.

48) Each Ovary of flowers produces

a) Only one seed.
b) Only two seed.
c) More than two seed.
d) Not produce any seed.

49) Ginger propagule in the form of

a) Runner.
b) Bulbils.
c) Bulb.
d) Rhizome.

50) Growth of pollen tube towards the ovule is called

a) Geotropism.
b) Chemotropism.
c) Phototropism.
d) Thermo tropism.

51) Red carpet welcome is happen in

a) Fallopian tube.
b) Uterus.
c) Cervix.
d) Vagina.

52) Function of seminal tube is

a) Provide fluid to move the sperm.
b) Nourishment of sperm.
c) Ejaculation of sperm.
d) All of the above.

53) Which of the following dieses is not a STD

a) Hepatitis B.
b) Herpes.
c) Syphilis.
d) None of the above.

54) Gametes are usually

a) Diploid.
b) Haploid.
c) Either (a) or (b).
d) None of these.

55) S curve is related to

a) Growth of human body.
b) Growth of plant body.
c) Growth of living organism.
d) Growth of sexual reproduction organs.

56) In India AIDS control organization is

a) NATO
b) NACO
c) NATA.
d) NACU

57) Either male of female reproductive organs are found in

a) Monoecious.
b) Dioecious.
c) Syngamy.
d) Meiocyte.

58) Grasses are pollinated by

a) Insects.
b) Water.
c) Wind.
d) Birds.

59) The sexual reproduction can be grouped into _____ distinct stages.

a) Two.
b) Three.
c) Four.
d) Five.

60) Germ cell of pollen converted to sperm cell by

a) Amitosis.
b) Mitosis.

c) Meiosis.
d) Germ cell are sperm cell.

61) Earthworm is one type of

a) Monoecious.
b) Dioecious.
c) Syngamy.
d) Meiocyte.

62) Full form of AIDS is

a) Acquired Immune Deficiency syndrome.
b) Actual Immune Deference syndrome.
c) Acquired Immediate Deficiency syndrome.
d) Anti Immune Deficiency syndrome.

63) Followings are the male sex hormones

1) FSH.
2) LH.
3) Progesterone.
4) Oestrogen.
5) GTH

a) 2 & 3.
b) 1, 2 & 4.
c) 1,2 & 5.
d) 1 & 5.

64) Sperms are matured and stored into

a) Epididymis
b) Spermatocyte.
c) Corpus lutetium.
d) Follicle.

65) Which virus is the cause behind AIDS

a) Ebola.
b) Tryponoema.
c) HIV.
d) Salmonella.

66) Followings are the female sex hormones

1) FSH.
2) LH.
3) Progesterone.
4) Oestrogen.
5) GTH

a) 2 & 3.
b) 1, 2 & 4.
c) 1,2 & 5.
d) All of the above.

67) Androecium is a group of

a) Style.
b) Female part of flower.
c) Stamens.
d) Petals.

68) Menstrual cycle varies between

a) 21-31 days.
b) 15-20 days.
c) 30-40 days.
d) 10-15 days.

69) Every pollen produces

a) 1 sperm cell.
b) 2 sperm cells.
c) 3 sperm cells.
d) 4 sperm cells.

70) Syphilis is one type of

a) STD.
b) Genetically Transmitted dieses.
c) Environmental dieses.
d) Mental dieses.

71) First stage of development of multi cell organism after fertilization is called

a) Cleavage.
b) Morella.
c) Blastula.
d) Gastrula.

72) Urino genital is found in

a) Female human body.
b) Earth worm.
c) Male human body.
d) None.

73) Which of the following is not the function of fertilization?

a) Transmission of genes.
b) Restoration of diploid chromosome.
c) Initiation of production of offspring's.
d) Production of placenta.

74) Every pollen are consists of _____ cells which comprise the male gametophyte.

a) 2
b) 2 or 3
c) 4.
d) Only 1.

75) Pollens are produced by

a) Amitoses.
b) Mitoses.
c) Meioses.
d) None of the above.

76) Gynoecium is

a) Style.
b) Female part of flower.
c) Stamens.
d) Petals.

[Answers & Explanations]

1) **(a).** Mitosis is related to body cell division and mutation is related to evolution.

2) **(c).** Those animals which lay eggs after fertilization inside the body are called Oviparous. Ex. Hen, Peahen, Gecko etc.

3) **(b).** Zygote formed after fertilization soit is diploid cell.

4) **(a).** Meiosis occurs to produce haploid cell sperm of ova. Both sperm and ova are called gametes.

5) **(a).** Ova is a female gamete cell and it's a haploid cell. Human female diploid cell contain 44 somatic chromosome and 2sex chromosome(XX type). So each ova will contain 22 somatic chromosome and 1 sex chromosome (X type)

6) **(c).** Human male diploid cell contains 44 somatic chromosome and 2sex chromosome(XY type). So each sperm will contain 22 somatic chromosome and 1 sex chromosome but 50% will have X type sex chromosome and 50% will have Y types sex chromosome.

7) **(a).** Those animals which gives birth of young one is called viviparous. Ex. Human being, Goat, Whale etc.

8) **(a).**

9) **(c).** 44 (22 pair) somatic chromosome and 2(1 pair) sex chromosome.

10) **(b).** Each human being have 2 sex chromosome. Male human cell have XY type sex chromosome pair and human female has XX type sex chromosome pair.

11) **(a).** Pollination—Fertilization—Seedling—Embryo.

12) **(b).** The organism behind malaria is *Plasmodium falciparum* and it reproduces by multiple fission.

13) **(a).** Father i.e. male cell have 2 different sex chromosomes(X and Y) but female sex chromosomes have same type(X) sex chromosomes so father pay the main role for determination of foetus sex.

14) **(c).** Similarities and variations both are transmitted.

15) **(a).** Scion part of a grafted tree normally superior in quality and that produces stem also.

16) **(a).** Water is the transfer medium for gametes of algae.

17) **(c).** Leishmania is the organism behind Kala-azar and its reproduces by binary fission.

18) **(b).** Uterus accept the eggs which are fertilized inside oviducts.

19) **(c).** Cell division is takes place in amitosis method.

20) **(a).** Scion normally selected superior in characteristics than stock during grafting.

21) **(b).** Scrotum is located outside the male body which is a pouch like body form and having prostate gland inside it. This is outside the body so that sperm could stay at lower temperature than body temperature.

22) **(b).**

23) **(c).** See also Q-11.

24) **(a).** Buds produced in leaves become an independent plant.

25) **(d).** Gametes—Polination—Fertilization—Zygotes—seedling—Cleavages—Embryo is the complete sexual reproduction steps of higher level plants.

26) **(b).** Male gametes i.e. sperms could move so they are called motile.

[A COMPLETE GUIDE TO MCQ]

CHAPTER: 08

27) **(a).**

28) **(d).** Sexual reproduction improve the gene quality and also helps to evaluate so it is the best mode of reproduction.

29) **(b).** During Tubectomy the fallopian tubes of female body are ligated so that ova could not come to fallopian tube and fertilization will stop.

30) **(b).**

31) **(b).** Unisexual flowers could produce food. They fertilizing by cross-pollination method.

32) **(a).** Placenta is the main source of nutrients for foetus.

33) **(a).** During Vasectomy in male body, vas deferens is ligated so that sperm could not reach outside the body to fertilize ova in female fallopian tube during intercourse.

34) **(b).** Asexual reproduction unit is spore.

35) **(c).** Ovary converted to fruits of a flower. Petals and stigma remains out of the fruit. Ovule converted to seeds.

36) **(a).** The giver options are the steps of fetus formation after fertilization. (a) to (d) are steps.

37) **(c).** IUCD is inter Uterus Contraceptive Device. Copper-T is inserted inside the uterus and stay there for 5-10 years as per wish of female. Copper ion helps to protects sperm for fertilization.

38) **(b).** See Q-34

39) **(b).** Reproduction rate is maximum for asexual reproduction method.

40) **(a).** IUCD means Intra Uterus Contraceptive Device. Copper-T, Hormonal-IUD are the example.

41) **(c).** See Q-21 for reason.

42) **(c).** It is related to unusual growth of prostate gland.

43) **(a).** See Q-33.

44) **(b).** By Insects: Lotus, Tulip etc. By water: water lily, algae etc. By Birds: Banyan tree, cherry etc. By Human: Artificial pollination(also called mechanical pollination)

45) **(b)** Such example found in Bryophyllum, Hydra etc.

46) **(b)** Prostate gland are produce prostate fluid which nourish the sperm cells.

47) **(d).**

48) **(a).**

49) **(d).**

50) **(b).** Chemotropism is the movement of body or body parts due to stimulation of chemicals.

51) **(b).** After fertilization in fallopian tube, the fertilized egg moves towards the uterus and blood vessels of uterus lock the movement of that egg. This is called red carpet welcome. After wards, egg started cell division and it become foetus to a complete young one.

52) **(a).** seminal fluid is lubricate the path of sperm up to ova and also helps to reduce friction during intercourse.

53) **(d).** All the given dieses are STDs(sexually transmitted dieses)

54) **(b).** Due to Meiosis cell division diploid cell become haploid gametes.

55) **(a).** Time-Vs-Growth curve of animal beings are generally found S in shape. This is called S-curve of growth,

[A COMPLETE GUIDE TO MCQ]

CHAPTER: 08

56) (b). NACO mean National AIDS control Organization.

57) (b).

58) (c).

59) (b). Gamete formation – fertilization- Development of fetus.

60) (b). Germ cells are converted to pollen by means of Mitoses method of cell division.

61) (a). Earthworms are one type of Monoecious. They have both male and female reproductive organs.

62) (a). AIDS: Acquired Immune Deficiency Syndrome

63) (c). Progesterone and Oestrogen are purely female hormones.

64) (a). Epididymis is the storage shelf of matured sperms.

65) (c). HIV=Human Immune Virus.

66) (d). All the hormones given are related to female sex organs.

67) (c).

68) (a).

69) (b)

70) (a).

71) (a).

72) (c) Urino Genital means which is used for transferring urine and sperms both. In human male body same path (penis) is used to extract urine and also to ejaculate sperm in female body so its called Urino Genital.

73) (d).

74) (b).

75) (c).

76) (b).

Heredity & Evolution

1) **New species may be formed if**

   1) A significant change takes place on genes of germ cells.
   2) There is no change in genetic material.
   3) Mutation takes place.
   4) Chromosome number change in the gametes.

   a) 1 & 3
   b) 1, 3 & 4.
   c) 1 & 2.
   d) 1 only.

2) **Who is known as father of genetics of heredity**

   a) Mendeleev.
   b) Gregor Johan Mendel.
   c) Hugo De. Vries.
   d) Charles Darwin.

3) **First law of Mendel is also known as**

   a) Law of segregation.
   b) Law of cross fertilization.
   c) Law of domination.
   d) Law of inheritance.

4) **Which of the following body characteristics describe as continuous variation**

   a) Height.
   b) Weight.
   c) Hair color.
   d) Eye color.

5) **A cat with white fur and blue eyes cross breed with another cat with same traits then in the subsequent generation**

   a) Cat with white fur will never have blue eyes.
   b) Cat with white fur will always have blue eyes.
   c) Cat with white fur may or may not have blue eyes.
   d) Cat with white fur will always have brown eyes.

6) **23rd pair of human male chromosome is**

   a) XX chromosomes.
   b) XY chromosomes.
   c) YY chromosomes.
   d) YX chromosomes.

7) **Which statement is correct.**

   a) Dominant traits cannot be expressed in heterozygous condition.
   b) Dominant traits can only be expressed in homozygous condition.
   c) Recessive traits can only be expressed in heterozygous condition.
   d) Recessive traits can only be expressed in homozygous condition.

8) **If we know the blood group of**

Father(O), mother(A) and daughter (O) of a family (members=4) then we could

a) Tell which group is dominant.
b) Not tell which group is dominant, as blood group of second child is not known.
c) Tell which group is dominant if we know the blood group of sisters and brothers of parents.
d) Not able to tell in any condition as its natural selection.

9) A _____ trait does not get expressed in F1 generation

a) Recessive
b) Dominant.
c) Pure.
d) None of the above.

10) When red flower plant is crossed breed with white flower and pink flower is produced in the next generation. Here pink character is

a) Dominant.
b) Recessive.
c) Co-dominant.
d) Epistatic.

11) Which one is correct about genetic drift?

a) Genetic drift does not involve completion between numbers of a species.
b) Genetic drift occurs in nature.
c) Genetic drift require presence of variations.
d) All are correct.

12) 23rd pair of human female chromosome is

a) XX chromosomes.
b) XY chromosomes.
c) YY chromosomes.
d) YX chromosomes.

13) Which of the following is heterozygous recessive

a) t.
b) TT
c) Tt
d) tt.

14) How many different types of gametes will be produced by a plant having genotype AABbCC

a) Two.
b) Three.
c) Four.
d) Nine.

15) Which branch of biology deals with transmission of characters from one generation to other

a) Heredity.
b) Evolution.
c) Genetics.
d) Biochemistry.

16) How many copies of gene(s) for a particular trait are present in germ cells

a) One.
b) Two.
c) Multiple.
d) Any number in different germ cells.

17) Which test is banned in throughout the world before birth of child

a) Prenatal growth rate determination.
b) Prenatal position determination.
c) Prenatal sex determination.
d) Prenatal blood group determination.

18) A person with unknown blood group suffered blood loss due to an accident. His friend came forward and offered blood donation. Doctor checked only the blood group of friend and started blood transfer. Friends blood group was

a) A

b) B
c) O
d) AB

19) The principle given by Darwin is

a) Inheritance of acquired characteristics.
b) Germplasm theory.
c) Theory of natural selection.
d) Mutation theory.

20) The principle given by Lamarck is

a) Inheritance of acquired characteristics.
b) Germplasm theory.
c) Theory of natural selection.
d) Mutation theory.

21) If a black cat is mates with a white cat and black color is dominant then F2 generation will have

a) 75% Grey cat, 25% White cat.
b) 75% Black cat, 25% Grey cat.
c) 75% Black cat, 25% White cat.
d) 100% Black cat.

22) Speciation is

a) Formation of a new hybrid species.
b) Formation of different physical traits.
c) Formation of new gene pool.
d) Convergence of evolution of different species.

23) Which principle given by Hugo de Varies

a) Inheritance of acquired characteristics.
b) Germplasm theory.
c) Theory of natural selection.
d) Mutation theory.

24) Excessive growth of hair on the pinna is a feature found

a) Only in Male body.
b) Only in Female body.
c) Both male and female body.
d) Mainly found in male body but sometime it also found in female body.

25) Excessive growth of hair on the pinna is a feature found only in male body because

a) The responsible gene is dominant in females but recessive in male.
b) The responsible gene is recessive in females but dominant in male.
c) This character is due to male testosterone hormone.
d) This character is suppressed in female body due to Oestrogen hormone.

26) Genetically dieses are more common in male compare to female because

a) Most dieses is due to X-linked recessive mutation.
b) Most dieses is due to Y-linked recessive mutation.
c) The dieses are X-linked dominant.
d) The dieses are Y-linked dominant.

27) Many of our physical characteristics like height, bone structure, hair and eye color are

a) Inherited.
b) Non inherited.
c) Secondary qualities.
d) None of the above.

28) In which animal the embryo develop into a male at high temperature

a) Turtle.
b) Pigeon
c) Grasshopper.
d) Fish.

29) Which one is correct

a) Recessive trait is present in phenotype but not appeared in genotype.
b) Recessive trait is present in genotype but not appeared in phenotype.
c) Recessive trait is present both in phenotype and in genotype.
d) Recessive trait never inherits.

30) In India, girl child is very less compared to boys because of?

1. Mother is responsible for girl child sex determination.
2. Killing of prenatal girl child.
3. Pre natal sex determination.
4. Poor health of girl child before birth.

a) All are correct.
b) 1,2 & 3 are correct.
c) 2 & 3 are correct.
d) 1 & 4 are correct.

31) If a black cat is mates with a white cat and black color is dominant then F1 generation will have

a) 75% Grey cat, 25% White cat.
b) 75% Black cat, 25% Grey cat.
c) 75% Black cat, 25% White cat.
d) 100% Black cat.

32) If two plants are crossed where one is tall and other is short then in F1 generation

a) 100% will be tall plant.
b) 50% will be tall plant and 50% will be neither tall and nor short plant.
c) 75% will be tall plant & 25% will be short plant.
d) 100% will be short plant.

33) If two plants are crossed where one is tall and other is short then in F2 generation

a) 100% will be tall plant.
b) 75% will be tall plant and 25% will be neither tall and nor short plant.
c) 75% will be tall plant & 25% will be short plant.
d) 100% will be medium size plant.

34) In DNA structure Guanine linked with

a) Cytosine.
b) Adenine.
c) Thymine.
d) Amino acids.

35) Alleles are segregated during

a) Gametes formation.
b) Fertilization.
c) Development of fetus.
d) Alleles are never segregated in any stage of life.

36) Archaeopteryx is the missing link between

a) Reptiles and birds.
b) Amphibians and reptiles.
c) Birds and mammals
d) Amphibians and birds.

37) In DNA structure Cytosine linked with

a) Guanine
b) Adenine.
c) Thymine.
d) Amino acids.

38) The branch of biology which related with heredity and variation is

a) Genetics.
b) Evolution.
c) Taxonomy.
d) Oncology.

39) Who gave Germplasm theory

a) Hugo de Vries.
b) Darwin.
c) Weismann
d) Lamarck

40) Missing link example is

a) Ostrich.
b) Dinosaurs.
c) Archaeopteryx
d) Gincobiloba

41) The wings of bat and wings of bird is an example of

a) Analogous organ.
b) Homologous organ.

c) Compound organ.
d) None of these.

42) Which animal was the common ancestor of human beings

a) Gorilla.
b) Chimpanzee.
c) Orangutan.
d) Lion tail Macau

43) Kale is a evolutionary product of

a) Wild cabbage.
b) Wild orange.
c) Wild pine.
d) Wild ass.

44) _____ is the naturally preserve organism as a prove of evolution

a) Living fossil.
b) Fossil.
c) Missing link.
d) All of the above.

45) Number of pairs of sex chromosome in zygote is

a) One.
b) Two.
c) Twenty three.
d) Twenty six.

46) Number of pairs of somatic chromosome in zygote is

a) One.
b) Two.
c) Twenty Three.
d) Twenty two.

47) Living animal fossil example is

a) Armadelo
b) Pangolin.
c) Owl.
d) Cockroach.

48) The layer of fossil gives the idea about

a) Existence period of the organism.
b) Reason of extinction.
c) Rate of evolution.
d) Degree of evolution.

49) Earliest member of human species is

a) Mammoth.
b) Apes.
c) Homo sapiens.
d) Eohippus.

50) Number of sex chromosome in zygote is

a) One.
b) Two.
c) Three.
d) Four.

51) Eohippus is the first generation of

a) Orangutans.
b) Ass.
c) Horse.
d) Gorilla.

52) The human hand and wings of pigeon is example of

a) Analogous organ.
b) Homologous organ.
c) Compound organ.
d) Evolutionary organ.

53) Living plant fossil example is

a) Gincobiloba.
b) Cactus.
c) Desert lili.
d) Dip nia.

54) A cross between a tall plant(TT) and short pea plant (tt)resulted in progeny that were all tall plants because

a) tt is dominant trait.
b) TT is dominant trait.
c) TT is recessive trait.

Heredity & Evolution MCQ (class X)

d) Height of pea plant is not governed by T or t.

55) Which of the following is correct regarding genes

1. Genes are specific sequence of bases in DNA molecule.
2. A gene does not code for protein.
3. Each chromosome has only one gene.
4. In individuals of a given species, a specific gene is located on a particular chromosome.

a) 1 & 4.
b) 1,2 & 3
c) 1,2 &4
d) All are correct.

56) Analogous organs are

a) Different structure and different function.
b) Different structure and same function.
c) same structure and different function.
d) same structure and same function.

57) Homologous organs are

a) Different structure and different function.
b) Different structure and same function.
c) same structure and different function.
d) same structure and same function.

58) Exchange of genetic material takes place in

a) Sexual propagation.
b) Asexual propagation.
c) Budding.
d) Vegetative propagation.

59) Which of the following statement is incorrect

1) For every hormone, there is a gene.
2) For every protein, there is a gene.
3) For production of every enzyme, there is a gene.
4) For every fat molecule, there is a gene.

a) 1 & 4.
b) 1,2 & 3
c) 2 &4
d) 4 only.

60) Two versions of a trait which are brought in by parents gametes are situated on

a) Homologous chromosomes.
b) Two different chromosomes.
c) Sex chromosomes.
d) Somatic chromosomes.

61) Alternating form of gene are called

a) Multiples.
b) Chromosomes.
c) Alleles.
d) Conjugate.

62) Inheritance of specific traits become clearer due to

a) Lamarck's theory.
b) Darwinism.
c) Mendel theory.
d) Neo-Darwinism.

63) Relation between sexual selection & natural selection is

a) Natural selection is a type of sexual selection.
b) Sexual selection is a type of natural selection.
c) Sexual selection occurs within demes but natural selection does not.
d) Sexual selection occurs during sex.

64) The term gene was coined by-

a) Watson and Crick.
b) Morgan.
c) Johannes
d) Horgovind Khorana.

65) Which of the following variations are not important from an evolutionary stand point.

a) Genetic differences between individual organisms comprising the populations.
b) Inherited differences between individual organisms comprising the populations.
c) Difference due to food, health, age that have no effect on the individual ability to survive & reproduce.
d) None of the above.

66) A X chromosome inherited from father will develop into a

a) Girl.
b) Boy.
c) X-chromosome does not determine the sex of the new born.
d) Either boy or girl.

67) Microevolution takes place due to

a) Somatogenic variation.
b) Blasto genic variation.
c) Successive variation
d) Continuous variation.

68) Forelimbs of vertebrates are example of

a) Analogous organs.
b) Homologous organs.
c) Vestibules organ.
d) Missing organs.

69) Which of the following is example of human vestibule organ

a) Ear muscle.
b) Abdominal muscle.
c) Canine.
d) All of the above.

70) A new species may formed if

a) DNA undergoes significant changes in germ cell.
b) Chromosome number changes in gametes.
c) There is no change in genetic material.
d) Gastrula.

71) In Mendel's peas, tall dominant to dwarf and yellow is dominant to green. A pure tall, yellow plant is crossed to a pure dwarf, green plant. What is the phenotype ratio in F2 generation?

a) 1:2:2:4:1:2:1:2:1:1.
b) 3:1
c) 1:2:1
d) 9:3:3:1

72) Vermiform appendix in man is an example of

a) Homologous organ.
b) Analogous organ.
c) Vestibule organ.
d) None of the above.

73) A representation of phenotype ratios of offspring is given by the

a) Independent assortment.
b) Law of product.
c) Punnett square.
d) Law of the sum.

74) By studying analogous structure basically, we are looking for

a) Similarities and appearance and functions but different in structure.
b) Similarities in structure.
c) Similarities in the cell make up.
d) Similarities in appearance but difference in functions.

75) The most evidence of evolutionary relationship is found in

a) Ocean beds.
b) Atmosphere.
c) Fossils.
d) Rocks.

76) Pairs of alleles that distribute randomly in gametes without regard to other pairs of alleles illustrate the

a) Independent assortment.
b) Law of product.
c) Law of segregation.
d) Law of sum.

77) Spontaneous generation of microorganism is disproved by

a) Charles Darwin.
b) Luis Pasteur.
c) Virchow.
d) Hugo de Varies.

78) The agreed concept on the development of first cell on earth is

a) On other planet.
b) In sea.
c) In rocks.
d) In soil.

79) How many gametes will be formed from TtYyRr

a) 8
b) 16
c) 20
d) 32

80) The flipper whale and the wings of birds are

a) Analogous organ.
b) Replaced organ,
c) Homologous organ.
d) Vestigial organ.

[A COMPLETE GUIDE TO MCQ]

CHAPTER: 09

[Answers & Explanations]

1) **(b).** Genetic material should change to develop new species. But this process is very very slow.

2) **(b).** He was a Christian priest.

3) **(a).** Law of segregation states that allele pairs separate or segregate during gamete formation, and randomly unite at fertilization

4) **(b).** Weight of the body of different species and individuals in the same species are varied randomly.

5) **(c).** White fur blue eyes WwBb

W for White fur, w for black fur, B for blue eyes, b for brown eyes.

M \ F	WB	Wb	wB	wb
WB	WWBB	WWBb	WwBB	WwBb
Wb	WWBb	**WWbb**	WwBb	**Wwbb**
wB	WwBB	WwBb	wwBB	wwBb
wb	WwBb	**Wwbb**	wwBb	wwbb

Here Cat with White fur may have Blue eyes (italic) or may not have blue eyes(bold)

6) **(b).** Human body have 22 pair of somatic chromosome and 23rd pair is sex chromosome. For male 23rd chromosome is XY and for female it is XX.

7) **(d).** Recessive traits means the traits which are not active. They only could appear in phenotype is the zygote is homozygote type. Example wwbb in Q-5

8) **(c).**

9) **(a).** See Q-7

10) **(d).**

11) **(a).** Genetic drift means variation in the relative frequency of different genotypes in a small population, owing to the chance disappearance of particular genes as individuals die or do not reproduce

12) **(a).** See Q-6

13) **(c).** Tt is heterozygous. TT and tt are homozygous.

14) **(c).** A, B, b and C

15) **(a).** The transformation of characteristic code from one generation to next generation is called heredity.

16) **(a).**

17) **(c).** To protect sex ratio.

18) **(c).** Type O blood is the called the universal donor because it has neither A nor B surface antigens on the red blood cell.

19) **(c).** Natural selection is the differential survival and reproduction of individuals due to differences in phenotype

20) **(a).** The theory of inheritance of acquired characteristics is If an organism changes during life in order to adapt to its environment, those changes are passed on to its offspring

21) **(c).** Let B for Black and b for White
In F2 progeny:

M \ F	B	b
B	BB	Bb
b	Bb	bb

[A COMPLETE GUIDE TO MCQ]   CHAPTER: 09

Here except bb all are a black cat. So the ratio will be Black: White:: 3: 1

22) **(c).** Speciation is the formation of the new gene pool.

23) **(d). Mutation theory**: A new species are formed from the sudden and unexpected emergence of alterations in their defining traits

24) **(a).**

25) **(b).**

26) **(a).** Due to XY shape, the male sex chromosomes are very unstable. Most genetic dieses are due to X-link recessive mutation.

27) **(a).**

28) **(a).**

29) **(b).** this traits are not appeared in phenotype that is why these are called recessive traits.

30) **(b).** Killing of prenatal girl child takes place due illegal use of sex determination method of ultrasonic machines and poor health of girl child because of negligence are the main cause of abnormally less sex ratio in India.

31) **(d).** In F1 progeny all cat color should be black due o its dominant nature.

32) **(a).** tall traits are dominant in nature so in F1 progeny all the plants will be tall only.

33) **(c).** In F2 progeny 75% will be tall and 25% will be short as a tall trait is dominant and short trait is recessive in nature. Also, refer example in Q-21

34) **(b).** In DNA adenine is linked with Guanine and Cytosine linked with Thymine.

35) **(a).**

36) **(a).** Archaeopteryx had both reptile and birds characteristics. It had beak, father, wings as birds and three chambered heart, cold blood as per reptiles.

37) **(c).**

38) **(a).** Genetically changes are the cause behind the variation.

39) **(c)** Hugo De Vries: Mutation theory. Darwin: Theory of natural selection. Weismann: Germplasm theory. Lamarck: Theory of inheritance.

40) **(c).** Archaeopteryx is the evolution proving a link between reptile and aves.

41) **(a)** Wings of Bat and birds having the same function but different structure so they are analogous organs.

42) **(b).**

43) **(a).**

44) **(d).** All are the evidence of evolution.

45) **(a)** In Zygote sex chromosome of father and mother is present and its form a pair.

46) **(d).** Somatic chromosome in human beings is 22 pair and its same in zygote.

47) **(d). Cockroach** present in earth since ancient time without any change so they are called living fossil.

48) **(a).** layers count indicates the age of the fossil.

49) **(c).**

50) **(b).** One pair. See Q-45.

51) **(c).** latest horse is Equus.

52) **(b).** As both the organs having the same structure but different function so they are called *Homologous* organ.

53) (a). *Gincobiloba* is called living plant fossil as it does not changed since ancient time of its existence in earth.

54) (b).

55) (a). Gene code for all

56) (b).

57) (c).

58) (a).

59) (d). For all, there is a gene.

60) (a).

61) (c).

62) (c).

63) (b). Sexual selection is a type of natural selection.

64) (a). They received Nobel prize for this in 1962.

65) (c).

66) (a). Girl have XX sex chromosome in which one from mother(XX) and another from father(XY).

67) (b).

68) (b). See also Q-52

69) (d). All the given organs are vestibule organs of human beings.

70) (a). Characteristics of a species are coded in genes.

71) (d). See Q-5 chart for comparing.

72) (c).

73) (c). The square drawn for answering Q-5 is an example of punnett square.

74) (a).

75) (c).

76) (a).

77) (a). By means of pasteurization method.

78) (b). Sea had the all required factors for life.

79) (a). TY, TR, Ty, Tr, tY. tR, ty, tr.

80) (c). Both having same structure but different functions.

Light Reflection & Refraction

1) The incident ray and reflected rays will be parallel for all angle of incident when two plane mirror are at an angle of

   a) 30°.
   b) 60°.
   c) 90°.
   d) 120°.

2) The focal length of a plane mirror is

   a) Zero.
   b) At infinity.
   c) Negative.
   d) Cannot be calculated.

3) In space speed of light is

   a) 3×10^{-8} m/sec.
   b) 3×10^{8} m/sec.
   c) 3×10^{10} m/sec.
   d) 3×10^{8} Km/sec.

4) If a convex mirror is fully submerged in water then the focal length of the mirror

   a) Will increase.
   b) Will decrease.
   c) Will remain same.
   d) Either increase or decrease and it depends on the density of water.

5) A convex lens can be regarded as a set of prisms and a glass slab, such that refracting angel of the prisms

   a) Continuously decreases in an outward direction.
   b) Continuously increases in an outward direction.
   c) Remain same in outward direction.
   d) None of the above.

6) For a spherical mirror, the relation between focal length and radius of curvature is

   a) $f = R$
   b) $f = 2R$
   c) $f = R/2$
   d) $fR = 2$

7) A convex lens is made of two sphere each of diameter 60 cm. Its focal length will be

   a) 30 cm.
   b) 120 cm.
   c) 60 cm.
   d) 15 cm.

8) The image formed by a concave mirror

a) Is always real.
b) Is always virtual.
c) Either real or virtual.
d) None of the above.

9) Power of a concave lens with f=20 Cm is

a) 4 D.
b) 4.5 D.
c) 6 D.
d) 5 D.

10) A concave mirror is a part of a sphere of diameter 120 cm. Its focal length will be

a) 30 cm.
b) 60 cm.
c) 120 cm.
d) 15 cm.

11) Velocity of light in a medium is 75% the velocity of light in vacuum then the refractive index of that medium is

a) 1.33
b) 0.75
c) 1.75
d) 2.52

12) The ratio of the size of the image to the size of the object is called

a) Power.
b) Focal length.
c) Magnification.
d) Transformation ratio.

13) A convex lens is made of two spheres each of diameter 30 cm. Its power will be

a) 1.33 D.
b) 2.33 D.
c) 3.33 D.
d) 13.33 D.

14) The number of images of an object held between two parallel mirror is

a) One.
b) Infinity.
c) Finite.
d) <10

15) If incident angle is 90 deg then Sine of refractive angle related to refractive index as

a) Sin r =µ
b) Sin r =1/µ.
c) Sin r = µ/2.
d) Sin r = µ^2.

16) Unit of lens power is

a) Debye.
b) Dioptre.
c) Decca.
d) Dalton.

17) Speed of light is maximum in following medium

a) Water.
b) Diamond.
c) Glycerin.
d) Air.

18) Arrange the light speed in increasing order

a) Water>air>diamond>glass.
b) Air>water>glass>diamond.
c) Diamond>glass>water>air.
d) Air>diamond>water>glass.

19) In angle of incident in medium, A is 60° and angle of refraction in medium B is 30° then the refractive index of medium B with respect to medium A is

a) 1.732
b) 1.23
c) 0.577
d) 1.52

20) The driver mirror used in an automobile is

a) Concave mirror.
b) Convex mirror.
c) Plane mirror.
d) Parabolic mirror.

21) If the critical angle of a substance is 30° then its refractive index will be

a) 2.
b) 0.5
c) 1.2
d) 1.33

22) A ray of light is incident on a concave mirror along its principle axis then what will be the angle of reflection?

a) 90°.
b) 180°.
c) 0°.
d) 270°.

23) The image from a convex mirror is

a) Always be real.
b) Always be projected.
c) Always be virtual.
d) Never be virtual.

24) In angle of incident in medium A is 60° and angle of refraction in medium B is 30° then the refractive index of medium B with respect to medium A is

a) 1.732
b) 1.23
c) 1.52
d) 0.577

25) If magnification factor is 2 and image size is 4 cm then what will be the actual size of the object?

a) 8 cm.
b) 6 cm.
c) 2 cm.
d) None of these.

26) What is the deviation produced by reflection at a plain mirror when the angle between the incident and reflected rays is 60°.

a) 210°.
b) 120°.
c) 30°.
d) 300°.

27) What will be the focal length of the concave mirror if the object is located 2.0 m from the concave mirror and image is 4.0 m away from the mirror

a) -2.0 m.
b) -0.67 m.
c) -1.67 m.
d) -1.30 m.

28) A real image formed by a convex lens is always

a) On the same side of the lens as the object.
b) Erect.
c) Inverted.
d) Smaller than the object.

29) An object is placed at the distance of 30cm from a concave mirror of focal length 15cm. The image will be

a) Real and of the same size.
b) Real and magnified.
c) Real and diminished.
d) Virtual and magnified.

30) If magnification factor is 2 and image size is 4 cm then what will be the distance of the object from the lens?

a) 8 cm.
b) 6 cm.
c) 2 cm.
d) None of these.

31) Which one is located behind the convex mirror?

a) The focal point.
b) A ray.
c) The real image.
d) The object.

32) For a plane mirror magnification factor is

a) 1
b) >1
c) <1
d) 0

33) A real and inverted image of the same size is formed by a concave mirror when the object is placed

a) Between the mirror and its focus.
b) Between the focus and center of curvature.
c) At the center of curvature.
d) Beyond the center of curvature.

34) The relation of focal length (cm) and power (D) of the lens is

a) $f = 1/P$
b) $f = P$
c) $f = 100/P$
d) $f = 2P$

35) Which lens will have more convergent rays

a) Lens having f = 15 cm.
b) Lens having f = 5 cm.
c) Lens having f = 25 cm.
d) Lens having f = -15 cm.

36) Refractive index indicates

a) The optical density of the material.
b) The actual density of the material.
c) The relative density of the material.
d) None of the above.

37) If the refractive index of diamond with respect to glass is 1.6 and absolute refractive index of glass is 1.5 then the absolute refractive index of the diamond will be

a) 2.4
b) 1.06
c) 0.94
d) 2.52

38) Absolute refractive index of any medium will be

a) 1
b) >1
c) <1
d) Infinity.

39) The power of plain mirror is

a) Infinity.
b) Zero.
c) Always negative.
d) Always positive.

40) What will be the focal length of a convex lens of power +2.5D

a) 40 cm.
b) 125 cm.
c) - 40 cm.
d) 0.4 cm.

41) The power of plain glass plate is

a) Infinity.
b) Zero.
c) Always negative.
d) Always positive.

42) The image formed by a plane mirror is

a) Real.
b) Real and erect.
c) Real and of the same size.
d) Virtual and laterally inverted.

43) Refractive index is maximum for

a) Glycerin.
b) Flint glass.

c) Diamond.
d) Vacuum.

44) Refractive index is minimum for

a) Glycerin.
b) Flint glass.
c) Diamond.
d) Vacuum.

45) If α is the angle of incident and β is the angle of refraction when light passed through water from the air then

a) α = β.
b) α > β.
c) α < β.
d) α ~ β.

46) If the image size is larger than object actual size then

a) Image distance is less than object distance.
b) Image distance is equal to object distance.
c) Image distance is larger than object distance.
d) It depend on the lens or mirror type.

47) If the focal length is made double then its power will

a) Doubled.
b) Halved.
c) Quadrupled.
d) Remains constant.

48) A real image formed by a convex lens is always

a) On the same side of the lens as the object.
b) Erect.
c) Inverted.
d) Smaller than the object.

49) Mirror formula is

a) 1/v + 1/f = 1/u
b) 1/v - 1/f = 1/u
c) 1/v + 1/u = 1/f
d) 1/f + 1/u = 1/v

50) lens formula is

a) 1/v + 1/f = 1/u
b) 1/v - 1/u = 1/f
c) 1/v + 1/u = 1/f
d) 1/f - 1/u = 1/v

51) Magnification formula of the lens is

a) v/u
b) – v/u
c) –u/v
d) u/v

52) Magnification formula of the mirror is

a) v/u
b) – v/u
c) –u/v
d) u/v

53) If an object is moved towards a convex lens, the size of the image

a) Decreases.
b) Increases.
c) First, decrease and then decrease.
d) Remains the same.

54) Second focal point is applicable for

a) Mirror only.
b) Lens only.
c) Both mirror and lens.
d) Neither mirror and nor lens.

55) 1D is

a) $1m^{-1}$
b) $1m$
c) $1cm^{-1}$
d) $1cm$

56) During refraction, the ray path changes due to

a) Different optical densities of the medium.
b) Incident angle >0°
c) Both (a) and (b).
d) None of the above.

57) If magnification factor appears as 0.2 then

a) Object distance > Image distance.
b) Object distance < Image distance.
c) Object distance = Image distance.
d) None of the above.

58) Refractive index with respect to air is same for

a) Balsam gum and glass.
b) Air and Water.
c) Diamond and Balsam gum.
d) Bothe (a) and (b) are correct.

59) A very small LED is placed at the focus of a convex lens, then the refracted beam will be a

a) A divergent beam of light.
b) A parallel beam of light.
c) A convergent beam of light.
d) A diffuse beam of light.

60) Refractive index is

a) c/v
b) v/c
c) cv
d) c/2v

61) Which one is correct?

a) Optical denser medium always physically has high mass density.
b) Optical denser medium is not always physically have high mass density
c) Optical denser medium refractive index is high.
d) Both (b) and (c).

62) The ratio of the sin of the angle of incident to sin of the angle of refraction is called

a) Law of polarization.
b) Law of reflection.
c) Law of dfraction.
d) Snell's law.

63) The ratio of the speed of light in a vacuum to speed of light in the medium is called

a) Refractive index.
b) Abs. refractive index.
c) Critical refractive index.
d) Mach number.

64) Bend of light rays in the denser medium is called

a) Reflection.
b) Refraction.
c) Diffraction.
d) Scattering.

[A COMPLETE GUIDE TO MCQ] CHAPTER:10

[Answers & Explanations]

1) **(c)**.

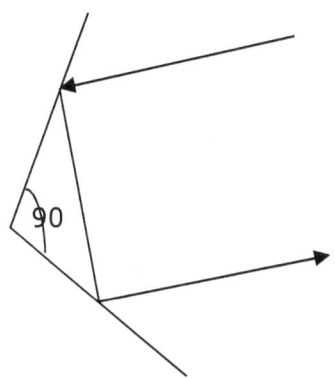

2) **(b). f =R/2** For plain mirror its R=infinity so the focal length is also infinity.

3) **(b)**.

4) **(c)**. Focal length will remain same as it is fully submerged.

5) **(b)**.

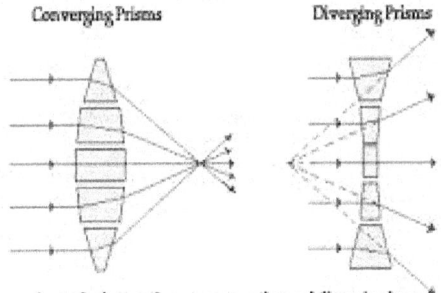

A set of prisms acting as a converging and diverging lens.

6) **(b)**.

7) **(d)**. focal length =R/2

   Given diameter=60 Cm.

   Radius = 30 Cm.

   Focal length = 30/2 =15 Cm

8) **(b)**.

9) **(d)**. Power of lens = (100/f) where f in Centimeter.
   Given f=20 Cm. Power = 100/20= 5 D.

10) **(b)**. focal length =R/2

    Given diameter=120 Cm.

    Radius = 60 Cm.

    Focal length = 60/2 =30 Cm

11) **(a)**. Refractive index = 1/75%

    =1/(3/4)

    =4/3 = 1.33

12) **(c)**. Magnification is the ratio of the size of the image to the size of the object For plane mirror it is 1.

13) **(d)**. focal length =R/2 & Power = 100/ f where f is in Centimeter.

    Given diameter=30 Cm.

    Radius = 15 Cm.

    Focal length = 15/2 Cm

    Power = 100×2/15 =13.33

14) **(b)**. For parallel mirrors, every image of one mirror is the object for another mirror. So there will be an infinite number of images.

15) **(b)**. μ = Sin i /Sin r

    When incident angle =90°

    Sin90° =1

    So, μ = 1 /Sin r

    Or, Sin r =1/μ.

Light Reflection & Refraction MCQ (class X)

16) (b).

17) (d). The speed of light is inversely proportional to an optical density which depends on refractive index. Air refractive index is lowest along the given options. So the velocity of light will be maximum in Air.
Here is the Refractive Index:

Air µ = 1.0003
Water µ =1.33
Glycerin µ= 1.32
Diamond µ=2.42

18) (b). For the same reason as define in Q-17. Glass µ = 1.52

19) (a). µ = Sin i /Sin r

µ = Sin 60°/ Sin 30°

µ = (√3/2) / (1/2)

µ = √3 = 1.732

20) (b).

21) (a). µ = Sin i /Sin r

When incident angle =90°

Sin90° =1

So, µ = 1 /Sin r

µ = 1/ Sin 30°

µ = 1/ (1/2)

µ =2

22) (c). Along with principle axis, it will follow the same path so the angle of reflection is 0°

23) (c).

24) (d). $^B\mu_A$ = Sin 30°/ Sin 60°

$^B\mu_A$ = (1/2)/ (√3/2)

$^B\mu_A$ = 1/√3 = 0.577

25) (c). Magnification = Size of the image/Size of the object.

m=2 and Image size is 4 Cm.

Object size =4/2=2 Cm.

26) (b). Angle of deviation = 180° – 60° =120°

27) (d). Here v= 4 Cm, u=2 Cm and f=?

Mirror equation 1/v +1/u = 1/f

¼ + ½ = 1/-f

1/-f = ¾

f= -4/3 =-1.33

28) (c).

29) (a) Here the object is located at a 2f distance so the image will be real and inverted. Only when the object is located <f distance for concave mirror its image becomes virtual and erect.

30) (c). m =2.

We know magnification = image distance/ object distance
=Image size /Object size.

So, Object size =Object distance=4/2=2 Cm.

31) (a).

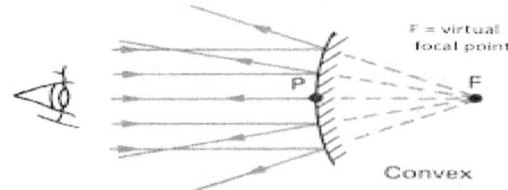

Reflection off a convex mirror

32) (a). Plane mirror magnification =1. As in this case always object distance is equal to image distance.

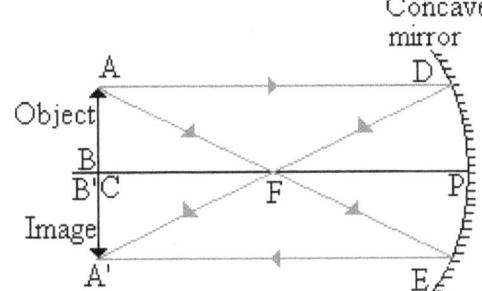

33) (c).

34) (c). When the focal length is in Cm the power is P=100/f.

So f =100/P

35) (b). Power is the ability of the lens to converge(Convex) or diverge (Concave) the rays after refraction through the lens. Power is inversely proportional to focal length. So lens with f =5 Cm will have the maximum power within the given options. So this lens will have more convergent rays.

36) (a). Refractive index indicates the optical density respect to light.

37) (a). $^D\mu_g$ =1.6 and $^g\mu_D$ =1.5

So absolute Refractive index of Diamond = 1.6 *1.5 =2.4

38) (b). Because light has maximum velocity in space and (abs.)μ =c/v

39) (b). Power = 1/ focal length.

For plain mirror its f= infinity.
So, power = 1/ infinity =0

40) (a). Power = 100/ focal length (when focal length in Cm).
Focal length = 100/2.5 =40cm

41) (b). See also Q-39.

42) (c).

43) (c). Refractive index:

Glycerin: 1.51, Flint glass: 1.65, Diamond:2.42, Vacuum medium=1

44) (d). See Q-43.

45) (b). In denser medium light rays bends towards the normal.

46) (c). See Q-30

47) (b). Power is inversely proportional to the Focal length. When focal length made doubled the power become halved.

48) (c).

49) (c).

50) (b).

51) (a).

52) (b).

53) (b).

54) (b).

55) (a). P=1/f. f in the meter. So D=m⁻¹

56) (c).

57) (a). See Q-30.

58) (d).

59) (b). The parallel beam of light with respect to the principle axis of the lens meet at the focus point. In opposite way, it is also correct.

60) (a).

61) (d). Optical density does not relate to mass density. it is related to the velocity of light through the medium.

CHAPTER: 10

62) (d). Snell's law: The ratio of the sines of the angles of incidence and refraction of a wave is constant when it passes between two given media. This constant is called refractive index.

63) (b). Absolute refractive index in the ratio of the speed of light and the speed of light through the medium.

64) (b).

HUMAN EYE & The colorful World

1) **Twinkling of stars is due to**
   a) Reflection of light.
   b) Diffraction of light.
   c) Refraction of light.
   d) Both (b) and (c).

2) **Observarory telescope contains**
   a) Concave lens.
   b) Convex lens.
   c) Plane mirror.
   d) Prism.

3) **Theoretically far point is located**
   a) At 1 km from the eye.
   b) At 25 cm from the eye.
   c) At infinity.
   d) At a finite distance but different for different people.

4) **Accommodation power of the eye is controlled by**
   a) Cornea.
   b) Pupil.
   c) Iris.
   d) Ciliary muscle.

5) **When light rays enter the eye, refraction of light occurs at**
   a) Iris.
   b) Pupil.
   c) Retina.
   d) Lens.

6) **Astigmatism can be rectified by**
   a) Progressive lens.
   b) Cylindrical lens.
   c) Convex lens.
   d) Concave lens.

7) **A progressive lens is used to rectify**
   a) Astigmatism.
   b) Presbyopia.
   c) Myopia.
   d) Hyper metropia.

8) **The vision of human eye could persist for(approx)**
   a) 0.1sec.
   b) 0.01sec.
   c) 1sec.
   d) 0.02sec.

9) **From the top of the sea its color is blue due to**
   a) Presence of phytoplankton.

b) Scattering of light.
c) Reflection of the sky.
d) Adsorption of light by the sea.

10) A person with long sight could not see_____ object clearly without _____ lens.

a) Near, Convex.
b) Distant, Convex.
c) Near, Concave.
d) Distant, Concave.

11) A person with short sight could not see_____ object clearly without _____ lens.

a) Near, Convex.
b) Distant, Convex.
c) Near, Concave.
d) Distant, Concave.

12) The intensity of light entered in the eye is checked by which sensory cells

a) Cornea.
b) Lens.
c) Rod cells.
d) Cone cells.

13) Near point of babies is

a) >25 cm.
b) <10 cm.
c) >100 cm.
d) =25 cm.

14) When ciliary muscle relax

a) Lens focal length is increased.
b) Lens focal length is decreased.
c) Lens focal lengths remain same.
d) Lens curvature of lens increases.

15) We can see Sun few minutes after sunset due to

a) Scattering of light.
b) Total internal reflection.
c) Dispersion of light.
d) Reflection of light.

16) Danger signal used Red color because

a) Scattering of red light is high.
b) Scattering of red light is less.
c) The velocity of red light is high.
d) It was in use since ancient time.

17) Which of the color in VIBGYOR is having least wave length

a) Violet.
b) Red.
c) Blue.
d) Yellow.

18) Which of the color in VIBGYOR is having a longest wave length

a) Violet.
b) Red.
c) Blue.
d) Yellow.

19) During sunrise and sunset the color of the sun appeared reddish due to

a) Scattering of light.
b) Dispersion of light.
c) Total internal reflection of light.
d) The polarization of light.

20) When ciliary muscle contract

a) Lens focal length is increased.
b) Lens focal length is decreased.
c) Lens focal lengths remain same.
d) Lens curvature of lens decreases.

21) Aero plane passenger flying at very high altitude will see the sky in

a) Red color.
b) Blue color.
c) White color.
d) Black(Dark) color.

22) The color of the cloud is mainly white due to

a) Presence of moisture.

b) Presence of dust particles.
c) Less scattering of blue light.
d) None of the above.

23) The clear sky appears blue due to

a) Absorption of blue light by air.
b) Violet and blue lights are scattered less than other color light.
c) Violet and blue lights are scattered more than other color light.
d) Due to UV ray.

24) The image formed by retina in the human eye is

a) Virtual and erect.
b) Real and inverted.
c) Real and erect.
d) Virtual and inverted.

25) Least distance for the distinct vision of the human eye is

a) 25 m.
b) 25 cm.
c) 125 cm.
d) Infinity.

26) The spectrum of sun light first proved by

a) Newton.
b) Euclid.
c) Hero.
d) Thomas Young.

27) In moon sky color is

a) Dark.
b) Blue.
c) Red.
d) White.

28) Which of the following produced due to scattering of light in the atmosphere

a) Tindal effect.
b) Polarization.
c) Interference.
d) Diffraction.

29) Which part of our eyes is responsible for accommodation power

a) Cornea.
b) Reina.
c) Cilliary muscle.
d) Optic nerve.

30) At noon sun appears white in color due to

a) Red color scatters most.
b) Blue color scatters most.
c) The heat of sun is less.
d) All colors of white scattered away.

31) For a person with _____ have far point less than infinity

a) Presbyopia.
b) Myopia.
c) Hyper Metropia.
d) Astigmatism.

32) For a person with _____ have near point more than 25 cm.

a) Presbyopia.
b) Myopia.
c) Hyper Metropia.
d) Astigmatism.

33) Primary colors are

a) Red,Yellow,Cyan.
b) Red,Cyan,Blue.
c) Red,Blue,Yellow.
d) Red,Green,Blue.

34) Following the sequence of phenomena is the cause behind the rain bow formation

a) Reflection, refraction and scattering.
b) Dispersion, scattering and total internal reflection.

The Human Eye & the Colorful World MCQ (class X)

c) Refraction, dispersion and total internal reflection.
d) Refraction, dispersion and internal reflection.

35) To rectify Presbyopia the lower part of the lens is consist of

a) Convex lens.
b) Concave lens.
c) Cylindrical lens.
d) Plano convex lens.

36) In moon sky color is dark due to

a) No presence of air particles.
b) No presence of the sun.
c) Scattering of light.
d) Dispersion of light.

37) The human eye lens is one kind of

a) Concave lens.
b) Convex lens.
c) Cylindrical lens.
d) Plano convex lens.

38) During sunset the shape of the sun is oval due to

a) Reflection of light by the earth.
b) Refraction of light.
c) Scattering of light.
d) All of the above.

39) The ability to focus eye at a different distance is called

a) Accommodation power.
b) Persistence of vision.
c) Near slightness.
d) Astigmatism.

40) The color of scattered light depends on

a) Size of the scattering particle.
b) The medium of colloid.
c) The velocity of scattering particle.
d) All of the above.

41) The color of fog light is

a) Red.
b) Yellow.
c) Blue.
d) Green.

42) If scattering medium particle size is very very less than the color of scattered light may

a) Red.
b) Blue.
c) Yellow.
d) White.

43) If scattering medium particle size is very very large then the color of scattered light may

a) Red.
b) Blue.
c) Yellow.
d) White.

44) If a person cannot see beyond 20cm. The person is having

a) Myopia.
b) Presbyopia.
c) Astigmatism.
d) Hyper metropia.

45) Hyper metropia is rectified by using

a) Cylindrical lens.
b) Convex lens.
c) Concave lens.
d) Plano convex lens.

46) The amount of light entering into the human eye is controlled by

a) Irish.
b) Pupil.
c) Retina.
d) Cornea.

47) Which of the following defects occurs in human eye lens due to age [main reason]

a) Presbyopia.
b) Myopia.
c) Hyper metropia.
d) Astigmatism.

48) Which of the following eye related defects could be rectified by using the cylindrical lens?

a) Presbyopia.
b) Myopia.
c) Hyper metropia.
d) Astigmatism.

49) Angel of deviation is smallest for which color in a spectrum when it passes through a prism

a) Red.
b) Blue.
c) Yellow.
d) Violet.

50) Myopia is rectified by using

a) Cylindrical lens.
b) Convex lens.
c) Concave lens.
d) Plano convex lens.

51) Which optical phenomena is used to produce cinema

a) Polarization.
b) Persistence of view.
c) Interference.
d) All of the above.

52) Size of the pupil is controlled by

a) Ciliary muscle.
b) Iris.
c) Aqueous humour.
d) Vitreous humour.

53) A spectrum produced by a prism with base angel 45°, 45° and vertex angel 90° when passing through an inverted _____ prism will produce a white light beam.

a) Base angel 45°, 45° and vertex angel 90°.
b) Base angel 60°, 60° and vertex angel 60°.
c) Base angel 30°, 60° and vertex angel 90°.
d) Base angel 60°, 70° and vertex angel 50°.

54) To rectify Presbyopia the upper part of the lens is consist of

a) Convex lens.
b) Concave lens.
c) Cylindrical lens.
d) Plano convex lens.

55) The red flower is seen red because

a) It absorbs red color light from white light.
b) It reflects red color light more than another color from white light.
c) It reflects only red color light from white light.
d) It absorbs white color light.

56) The red flower is seen black in blue color light because

a) It absorbs blue color light.
b) It reflects blue color light.
c) It does not reflect any light.
d) A person having color blindness.

57) A person with color blindness having problem

a) With Principle colors.
b) With red, green and blue colors.
c) Both (a) and (b)
d) Only with blue color.

58) Indigo is mixed with water during cleaning of white dresses

a) To clear yellowish patch.
b) To clear bluish patch.
c) To clear all color patch.
d) To sterilize the cloth.

59) A student has difficulty in reading the whiteboard while sitting in the last row because

a) Board is white.
b) The student has myopic eyes.
c) The student has hyper metropia eyes.
d) The student has Presbyopia.

60) Retina consists of

a) Rod cells only.
b) Cone cells.
c) Rods and cones cells.
d) None of the above.

61) The angle of deviation is minimum for

a) Red color light.
b) Yellow color light.
c) Blue color light.
d) Violet color light.

62) Planets do not twinkle because

a) No refraction takes place in the atmosphere.
b) Variation of the intensity of light is approximately zero.
c) All light absorbed by the atmosphere.
d) All light is reflecting back.

[A COMPLETE GUIDE TO MCQ]

CHAPTER: 11

[Answers & Explanations]

1) **(c).** Due to continuous refraction of light coming from the star the intensity of light changes continuously. This is the reason behind twinkling of stars.

2) **(b).** Convex lens.

3) **(c).** Theoretically human being could focus up to infinity.

4) **(d).** Ciliary muscle compress of expanding the lens to focus it

5) **(d).** Light rays refracted in the eye only through the lens.

6) **(b).** Astigmatism is a problem of the axis. Cylindrical lens used to rectifying his problem.

7) **(b).** Progressive lens is a modern form of a bifocal lens.

8) **(a).** Human eye vision could persist for 1/16 sec approximately.

9) **(c).** Raman effect also defines this.

10) **(a).** A person with long sight is called hyper metropia. They could not see near object clearly without a convex lens.

11) **(d).** A person with short sight called myopia. They could not see the distant object clearly without a concave lens.

12) **(c).** High intensity and color are sensed by cone cell.

13) **(b).** Near point of babies is 5-8 Cm.

14) **(a).** Due to relax state of ciliary muscle lens radius of curvature increases so lens focal length also increases. $f = R/2$

15) **(a).** Due to scattering of light

16) **(b).** Scattering of red light is less as its wave length is maximum. According to the theory of scattering its inversely proportional to wave length.

17) **(a).** Violet color wavelength is minimum and frequency is maximum in VIBGYOR.

18) **(b).** Red color wavelength is maximum and frequency is minimum in VIBGYOR.

19) **(a).** Red light scattering is minimum than another color so during sun set and sun rise only this color could reach up to our eyes but other colors are scattered on the way. Due to this they appeared reddish.

20) **(b).** During contract of ciliary muscle it decreases the radius of curvature of the eye lens so its focal length also decreases. $f = R/2$

21) **(d).** At very high altitude due to the absence of cloud and dust particles the sky will appear dark i.e. black color.

22) **(a).** Due to the presence of moisture light refracted and its color appears white.

23) **(c).** Violet and blue color scattered most compare to another color so the sky appears blue. The human eye is sensitive to blue color compared to violet so its blue only.

24) **(b).** The object is located more than the 2f distance in front of the convex lens so it should be always real and inverted.

25) **(b).** 25 Cm. For babies it is 5-8 Cm.

26) **(a).** Newton

27) **(a).** In Moon sky is not containing any cloud and dust particles as earth so due to no scattering of light it will appear dark or black in color.

[A COMPLETE GUIDE TO MCQ]   CHAPTER: 11

28) **(a).** Tindal effect is produced by the scattering of white light in presence of particles.

29) **(c).** Ciliary muscle by means of a contract and relax change the focal point of the human eye. This ability is called accommodation power of the eye.

30) **(d).** Due to vertical position all color of white ray equally scattered and its color appears white.

31) **(b).** A person with myopia could not see at a long distance so it's far point is less than infinity. As per theory human far point should be at infinity.

32) **(c).** A person with hyper metropia could not see properly at near sight but could see the far sight. This persons near sight are more than 25 Cm.

33) **(c).** Primary colors are those colors which could not be produced by mixing of other colors at any ratios. Red, Yellow and blue are a primary color and others are compound color.

34) **(c).**

35) **(a).** A person with Presbyopia uses the convex lens at the lower part of the eyeglass to see the near point objects.

36) **(a).**

37) **(b).**

38) **(b).** Due to refraction of light from low density medium to high density medium light rays change the path. As eye always view straight path so sun seems oval in shape during sunset when the light has to travel more path compared to noon time.

39) **(a).**

40) **(a).** According to particle size the scattered color changes.

41) **(b).** Yellow color light has the least angle of deviation.

42) **(b)** If the size of the particle is very very less
then possible color which it could be scattered is blue because its wave length is minimum. If the particle size is more then it became may be red.

43) **(a).** See Q-42.

44) **(a).** A person with myopia cannot have far point of view.

45) **(b).** Convex lens focuses the object at retina which was focused *after* the retina before wearing the convex lens.

46) **(b).** By maintaining the opening diameter of pupil it controls light entering into the human eye.

47) **(a).** Presbyopia mostly appears at age 40 years and above.

48) **(d).** It is the problem of the axis.

49) **(c).**

50) **(c).** Concave lens focuses the object at retina which was focused *before* the retina without wearing the concave lens.

51) **(b).** Cinema is the continuous appearance of pictures image in the eye. For human being brains image persistent time is 1/16 sec and for sound its persistent time is 1/10. So as they are approximately same this theory used to produce moving cinema.

52) **(b).** Iris' muscle controls the opening and closing of the pupil.

53) **(c).** Only same prism if inverted then the spectrum will again join and give a white color ray of light.

54) **(b)**

55) (c). the red flower is seen red because when white light fall on it only red color reflected from the flower and reach up to our eyes. All other colors are absorbed by the flower.

56) (c). Red color flower could reflect only red color. All other colors are absorbed by the flower. Due to the absence of color it will appear black.

57) (b). A person with color blindness having a problem to see fully Red, green and Blue.

58) (b). Generally dirty cloth having a yellow patch and the blue color is the compensatory color of yellow color so blue is used. Yellow + Blue = White.

59) (b). The student has myopia so he could not focus at retina by means of accommodation.

60) (C). The retina is consists of Rod and cone cells.

61) (b). The yellow color angle of deviation is minimum.

62) (d). Planets are very nearer to us compared to the sun. Planets are seen by their reflecting light which has very less intensity than stars. So the intensity of light does not change much in case of planets. This is the reason why planets are not twinkling as same as stars.

ELECTRICITY

1) **What is the unit of rate of charge flow in SI unit?**

   a) Siemens.
   b) Ampere.
   c) Volt.
   d) Watt.

2) **The voltage at the positive terminal of a battery compared to the negaive terminal is**

   a) Less due to less presence of electrons.
   b) More due to more presence of electrons.
   c) More due to less presence of electrons.
   d) Less due to more presence of electrons.

3) **A battery of 6 Volt is connected with resistance 2Ω & 4 Ω in series then what will be the power consumption due to circuit resistance**

   a) 6 watt.
   b) 7 watt.
   c) 36 watt.
   d) 16 watt.

4) **The relation between Potential difference V, Work done W & Charge Q is**

   a) V= Q ×W
   b) Q=V/W
   c) W=V×Q
   d) V=W/Q

5) **The current measuring instrument is**

   a) Potentiometer.
   b) Ammeter.
   c) Galvanometer.
   d) Voltmeter.

6) **The velocity of electrons through the conductors is**

   a) 3×10^7 m/s.
   b) 3×10^8 m/s.
   c) 3×10^9 m/s.
   d) 3×10^{10} m/s.

7) **1 volt = _____/ 1 Coulomb**

   a) 1 Jule
   b) 1 watt.
   c) 1 Ohm.
   d) 1 Newton.

8) **The function of an electric fuse is based on**

   a) The magneic effect of current.
   b) The inductive effect of current.
   c) The jule effect of current.
   d) The chemical effect of current.

9) **A component used to regulate current without changing the voltage source is called**

   a) Rheostat.
   b) Variable ammeter.
   c) Potentiometer.
   d) Meter bridge.

10) **When one unit of charge is moved from one point to another point in an electric field, the amount of work done in joule is called____ between the two points**

   a) Electric current.

Electricity MCQ (class X) [Page 110]

b) Electric power.
c) Electric potential.
d) Electric inductance.

11) Which of the following statement(s) is wrong regarding the electrical set up for verification of Ohm's law?

1) The voltmeter is connected in series with the known resistance.
2) The rheostat is used to increase the current by keeping resistance same.
3) A single key is used to switch ON and OFF the circuit.
4) An ammeter is connected in series with the resistance.

a) 1 & 3 are wrong.
b) 1 & 2 are wrong.
c) 3 & 4 are wrong.
d) Only 2 is wrong.

12) The potentiometer is used to measure

a) Potential of the cell.
b) The voltage of cell.
c) The resistance of cell.
d) The conductance of cell.

13) To define Ohms law following parameter should remain constant

a) Temperature.
b) Dimension.
c) Material.
d) All of the above.

14) When electric current pass through a conductor from the positive terminal to the negative terminal of the battery, the electrons are flow

a) From positive to negative terminal.
b) From negative to the positive terminal.
c) In any direction it could flow.
d) Some electrons flow from positive to negative and some in opposite direction to maintain the growth of electron at any point at any instant of time as zero.

15) Alloys are generally used for making of the heating coil due to

1) The resistivity of alloys is generally higher than constituent metals.
2) Alloying materials resist corrosion and oxidation at a high temperature.
3) Alloying material melting point temperature is more than constituent metal.

a) 1 & 2 are correct.
b) 2 & 3 are correct.
c) 1 & 3 are correct.
d) All are correct.

16) 1KWh =

a) 4.2×10^6 Joule
b) 3.2×10^6 Joule
c) 3.6×10^6 Joule
d) 1.8×10^6 Joule

17) An electric kettle consumes 1kW of electric power when operated at 220V. A fuse wire of what rating shall be used to protect it.

a) 1A.
b) 2A.
c) 4A.
d) 5A.

18) Domestic consumption of electric power is measured in

a) Watt-h.
b) B.O.T unit.
c) Ton.
d) Million Unit.

19) Resistivity is the property depends on

a) Material
b) Voltage source.
c) Dimension.
d) All of the above.

20) SI unit of resistivity is

a) Ohm-m

b) Siemens
c) Amp/m
d) Jule-m

21) SI unit of conductivity is

a) Ohm-m
b) Siemens/m
c) Amp/m
d) Jule-m

22) Electric bulb filament made of

a) Tungsten.
b) Nicrome.
c) Stainless steel.
d) Manganin.

23) Electric heater coil generally made of

a) Tungsten.
b) Nicrome.
c) Stainless steel.
d) Ebonite

24) A, B and C are three resistors whose resistance are in A.P. If the resistance of A is one fourth of resistance C where the resistance of C is 8 Ω, then find resultant resistance if they are connected in series.

a) 15 Ohm.
b) 18 Ohm.
c) 1.5 Ohm.
d) 25 Ohm.

25) In _____ connection current is same along all resistance

a) Parallel.
b) Series.
c) Mixed.
d) Delta.

26) In _____ connection equivalent resistance is less than the any of the circuit resistance.

a) Parallel.
b) Series.
c) Mixed.
d) Delta.

27) In _____ connection equivalent resistance is more than the any of the circuit resistance.

a) Parallel.
b) Series.
c) Mixed.
d) Delta.

28) In _____ connection voltage is same along all resistance

a) Parallel.
b) Series.
c) Mixed.
d) Delta.

29) Substances which have large numbers of free electrons and offer low resistance to electron flow are called

1) Pure metals.
2) Semi-conductors.
3) Thermistors.
4) Insulator.
5) Inductors.

a) 1,2 & 3 are correct.
b) 2,3 & 4 are correct.
c) 1 & 5 are correct.
d) 1 is correct.

30) Domestic wiring is an example of

a) Parallel connection.
b) Series connection.
c) Delta connection.
d) Star connection.

31) The ratio of mass of the proton to the electron is

a) 1480
b) 1835
c) 50
d) 4

[A COMPLETE GUIDE TO MCQ]   CHAPTER:12

32) Incandescent lamp generally filled with

a) Hydrogen gas.
b) Chlorine gas.
c) Helium gas.
d) Nitrogen gas.

33) The nicrome wire is an alloy of

a) Nickel + Vanadium.
b) Chromium + vanadium.
c) Nickel + Chromium.
d) Aluminum + Nickel

34) Following has zero temperature co-efficient

a) Liquid metals.
b) Electrolytes.
c) Dry salts.
d) Thermistor.

35) Fuse wire connected with the main circuit in

a) Parallel connection.
b) Series connection.
c) Delta connection.
d) Star connection.

36) Fuse wire generally made of

a) High resistance wire material.
b) Low resistance wire material.
c) High melting point temperature wire material.
d) Stainless steel wire.

37) The range of domestic line voltage is

a) 110 V to 120V.
b) 200 V to 220V.
c) 400 V to 420V.
d) 110 V to 220V.

38) Melting point temperature of fuse wire is

a) Very high.
b) Moderate.
c) Very low.
d) Always more than 500ºC.

39) According to o Joule law of heating

a) $H=I^2Rt$
b) $H=IR^2t$
c) $H=IRt$
d) $H=I^3Rt$

40) The temperature coefficient of electrolytes is

a) Positive with high value.
b) Positive with low value.
c) Negative with low value.
d) Nil.

41) Power =Voltage ×_____

a) Ampere.
b) Oham.
c) Coulomb.
d) Farad

42) A cylindrical conductor of length l and uniform area of cross section A has resistance R. Another conductor of length 2l and resistance R of the same material has an area of cross section

a) A/2.
b) 3A/2.
c) 2A.
d) 3A.

43) Ampere is_____ quantity.

a) Scalar.
b) Vector.
c) Neither scalar nor vector.
d) It has magnitude only.

44) An electric bulb is rated 220V and 100Watt.When it is operated on 110V , the power consumed will be

a) 100W.
b) 75W.
c) 50W.

Electricity MCQ (class X)

d) 250W.

45) How many electrons per second pass through a given point in a wire carrying 4Amp current?

a) 1.25×10^{19}
b) 2.5×10^{19}
c) 5×10^{19}
d) 3×10^{19}

46) A filament of an electric bulb draws a current of 2 Ampere. A number of an electron passing through a cross section of the filament in 16 seconds would be ?

a) About 2×10^{16}
b) About 2×10^{18}
c) About 2×10^{20}
d) About 2×10^{23}

47) Conductance :mho::

a) Resistance :Siemens
b) Inductance :Farad
c) Resistance :ohm
d) Capacitance :Hennery

48) If a conductor length is increased, 100% (remaining all other parameters fixed) then what will be the increase in power dissipate.

a) 100%.
b) 200%.
c) 300%.
d) 400%.

49) An electric iron has a rating of 800W,200V.What is its resistance?

a) 50Ω.
b) 160Ω.
c) 250Ω.
d) 150Ω.

50) In an electric circuit 2 resistors of 2Ω & 4Ω respectively are connected in series to a 6 V battery. The heat dissipated by a 4Ω resistor in 5 sec is

a) 5J.
b) 10J.
c) 20J.
d) 30J.

51) Van De Graff Generator generate

a) High static voltage.
b) High dynamic voltage.
c) Low current and high dynamic voltage.
d) High current and high static voltage.

52) Graph between voltage and currents for metals is a

a) Straight line.
b) Curve.
c) Parabola.
d) Sinusoidal.

53) Electric power over long distance is transmitted at

1) High voltage.
2) Low voltage.
3) High current.
4) Low current.

a) 1 & 3.
b) 2 & 3.
c) 1 & 4.
d) 2 & 4.

54) Electric fuse woks on the principle of

a) Joule heating.
b) Ohm law.
c) Ampere law.
d) None of the above.

55) A current of 1A is drawn by a filament of an electric bulb for 10 minutes. Find the amount of electric charge that flows through the circuit.

a) 300C
b) 150C
c) 600C
d) 60C

56) What will be the length of nicrome wire of resistance 5Ω, if the length of similar wire of 120 Cm has the resistance of 2.5Ω.

a) 240 Cm.
b) 24 Cm.
c) 48 Cm.
d) 480 Cm.

57) Semi-conductors have electrical conductivity

a) More than metals but less than insulators.
b) Less than thermistor but more than an insulator.
c) Equal to the thermistor.
d) Less than metals but more than typical insulators.

58) During short circuit resistance become

a) Infinity.
b) Zero.
c) Does not change.
d) Data insufficient.

59) When open resistor checked by Ohm-meter its resistance

a) Infinity.
b) Zero.
c) Does not change.
d) Data insufficient

60) The resistance of a conductor varies proportionately with

1) Area of the cross section.
2) Temperature.
3) Length.
4) Resistivity.

a) 1, 3 & 4.
b) 2 & 3.
c) 2, 3 & 4.
d) 1, 3 & 4.

61) The conductivity of semi-conductors will_____ with an increase in temperature.

a) Decreases.
b) Increases.
c) First increase & then decrease.
d) Remains constant.

62) When temperature increases the conductance of conductors is

a) Decreases.
b) Increases.
c) First increase & then decrease.
d) Remains constant.

63) Voltage dependent resistors are generally made from

a) Charcoal.
b) Silicon carbide.
c) Nicrome.
d) Copper wire.

64) Ohm law is applicable to

1. A.C circuit.
2. D.C circuit.
3. Semi-conductor.
4. Super conductors.

a) 1, 3 & 4.
b) 1 & 2.
c) 2, 3 & 4.
d) 1, 2 & 4.

65) Three resistor each having resistance 5 Ω is first connected in parallel and then in series, the ratio of the resultant resistance of the first combination with second is

a) 1/9.
b) 1/8.
c) 1/5.
d) 1/10.

66) Street lights are connected in

a) Series.
b) Parallel.
c) Either series or parallel.

d) End to end.

67) When current flows through heater coil it glows but supply wire does not glow because

a) The resistance of supply line is more than heater coil.
b) Supply wire covered with PVC.
c) The resistance of supply line is less than heater coil.
d) Resistance coil has nicrome as a metal.

68) When the load is switched off sparking occurs due to

a) The resistance of the circuit.
b) The inductance of the circuit.
c) The capacitance of circuit.
d) The impedance of the circuit.

69) International Ohm is defined in terms of the resistance of

a) A column of mercury.
b) A cube of carbon.
c) A cube of copper.
d) The unit length of wire.

69) If there are 3 resistances each of 2Ω and due to their combination effective resistance found as 3Ω, then how the resistances are connected

a) Three are in parallel.
b) Three are in series.
c) Two resistances are in parallel combination and another one is in series with that combination.
d) Two resistances are in series combination and another one is in Parallel with that combination.

70) Following apparatus is not needed for verification of Ohm's law

a) Galvanometer.
b) Ammeter.
c) Voltmeter.
d) Rheostat.

71) Which of the following in/are a conductor

1) Dry HCl.
2) Graphite.
3) Carbon black.
4) Rubber.

a) 1 is correct.
b) 2 is correct.
c) 2 & 3 are correct.
d) All are correct.

72) Which statement(s) are not correct

1) A voltmeter connected in series.
2) Ammeter connected in series.
3) Rheostat could be connected in series or parallel.
4) During parallel connection heat dissipation is more from the circuit.

a) 1 & 3.
b) 2 & 3.
c) 3 & 4.
d) 2 & 4.

73) When ohmmeter reads zero, its indicates the resistor is

a) Shorted.
b) Opened.
c) Instrument faulty.
d) The added resistor is an insulator only.

74) In parallel combination total equivalent resistance decreases due to

a) Increase in length.
b) The decrease in the area of cross section.
c) Increase in area of cross section.
d) Increase in the current.

75) The solar cell is made of ?

a) Semi-conductors.
b) Super conductors.
c) Insulators.
d) Conductors.

76) Which one is odd

a) Semi-conductor.
b) Thermistor.
c) Conductor.
d) Semi-insulator.

77) Fuse wire not coated with insulator because

a) It should dissipate heat during normal time.
b) It is connected in parallel to the circuit.
c) It requires to melt instantaneously.
d) None of the above.

78) In series connection of resistance equivalent resistance increases due to

a) Increase in length.
b) Increase in the cross sectional area.
c) Increase in temperature.
d) Same current along all resistances.

79) There are two bulbs i.e. 200W and 100W. They are connected in same voltage. Which valve having more resistance?

a) 200W
b) 100W
c) Both are same.
d) Data insufficient to calculate.

80) The ideal resistance of a Voltmeter should be

a) Infinity.
b) Zero.
c) 1000kΩ
d) Negligible.

81) The ideal resistance of an ammeter should be

a) Zero.
b) Infinity.
c) 1000kΩ
d) Negligible.

82) Electric bulbs consists of _____terminals

a) 1
b) 2
c) 3
d) 4

83) The industrial motor consists of _____ number of terminals.

a) 2
b) 3
c) 4
d) Either 2 or 3

[A COMPLETE GUIDE TO MCQ]

CHAPTER: 12

[Answers & Explanations]

1) **(b)**. The rate of charge flow is current and its unit is Ampere.

2) **(c)**. Positive terminal contains less electron than negative terminal. Electron flow from high voltage (positive terminal) to low voltage (negative terminal)

3) **(a)**. Equivalent resistance (Series connection)= 2 + 4 =6Ω

Current = 6V/6Ω =1Amp. Power =V×I =6 ×1=6Watt.

4) **(d)**. Voltage is work done required to displace per unit charge.

5) **(b)**. Ampere= Charge/time.

6) **(b)**. The velocity of electrons through a conductor is approximately equal to the velocity of light in space i.e. 3×10^8 m/S

7) **(a)**. Voltage is work done required to displace per unit charge.

8) **(c)**. Electric fuse function of joule effect of current. When excess current try to flow through the circuit it melted due to overheating and disconnect the circuit. It protects the circuit from damage.

9) **(a)**. The rheostat is an instrument, which used to change the resistance.

10) **(c)**. Its definition of voltage.

11) **(b)**. The rheostat is an instrument to change resistance and voltmeter always connected in parallel with the voltage source.

12) **(a)**. potentiometer is used to measuring open circuit voltage i.e. e.m.f and voltmeter is used to measure closed circuit voltage.

13) **(d)**. All the given factors change the resistance so all should be remain constant to define ohms law.

14) **(b)**. The direction of flow of electron is opposite to current flow direction.

15) **(d)**. All statements are correct.

16) **(c)**. 1kWh = 1000W × 1Hr
= 1000W × 3600 S
= 3600000 joule.
= 3.6×10^6 joule.

17) **(d)** Current = 1000/220 ~5A . Fuse wire should be rated for maximum current.

18) **(b)**. 1Kwh = 1 B.O.T(Board of trade unit)

19) **(a)**. Resistivity is the property of the material that determines the resistance between the two opposite surface of a unit cube at a constant temperature. All materials resistivity are different. the reciprocal of resistivity also called conductivity.

20) **(a)**. SI unit of resistivity is Ohm-meter.

21) **(b)**. SI unit of conductivity is Simens /meter. See also Q-19 for definition.

22) **(a)**. Tungsten melting point temperature is very high and its resistance also very high.

23) **(b)**. Nicrome is an alloy of Nickel and chromium. It has high melting point temperature and not react with air at high temperature.

24) **(a)**. C =8Ω So, A =8/4=2Ω

As A,B,C are in A.P

B = (A+C)/2 =(2+8) / 2 =5Ω

Total Resistance = 2+5+8=15 Ω

25) **(b)**. In series connection current is same for all the resistance but voltage defer.

Electricity MCQ (class X)

26) (a). In parallel connection equivalent resistance is less than the any of the circuit resistance as its equivalent resistance reciprocal is the sum of reciprocal of all the resistance of the given circuit.

27) (b). In series connection equivalent resistance is the sum of all resistance so equivalent resistance is always more than any resistance of the given circuit.

28) (b).

29) (d). Only pure metals have large numbers of free electrons and they offer low resistance to electron flow.

30) (a). Domestic wiring connection is always made parallel so that all electrical equipments could connect across same voltage and failure of one equipment will not affect other.

31) (c). Mass of proton(M_p): 1.67×10^{-27} Kg

Mass of electron(M_e): 9.11×10^{-31} Kg

Ratio: $M_p/M_e = \dfrac{1.67 \times 10^{-27} \text{Kg}}{9.11 \times 10^{-31} \text{Kg}} = 1833$

32) (d). Earlier incandescent bulbs are not filled with any gas. To reduce the vaporization of tungsten wire Nitrogen bulb later filled in it.

33) (c). Nickel + Chromium.

34) (d). Thermistor resistance remains constant with increase in temperature.

35) (b). Fuse wire is connected in series so that fault of the line could not transfer to the equipment and equipment will remain safe.

36) (b). Fuse wire generally made of low resistance wire so that due to the flow of high current it could produce high heat to melt and break the circuit.

Heat produced due to current is proportional to resistance but proportional to the *square* of the current flow. A small change in current flow will increase heat greater than resistance change. $H = I^2 R$.

37) (b). 200-220V is safe for domestic use.

38) (c). Melting point temperature of fuse wire should be very low so that it could melt easily due to the production of heat during the flow of fault current through the circuit.

39) (a). According to Joule heat law the heat produced due to the flow of current through a resistance is proportional to the resistance of the resistor, time of current flow and square of the current flowing in it.

Heat = $(\text{Current})^2 \times \text{Resistance} \times \text{time}$.

40) (c). The temperature coefficient of the electrolyte is negative as due to increasing in temperature number of electrons and its mobility increases.

41) (a). Power = Voltage × Ampere.

42) (c). $R_1 = \rho L/A$ and $R_2 = \rho 2L/x$

[let area of the cross section for the second resistor is x]

Now, $R_1 = R_2$

$\rho L/A = \rho 2L/x$

$X = 2A$

43) (a). Current unit has value only so it is a scalar quantity.

44) (c).

45) (b). $I = nq/\text{time}$

$4 = n \times 1.6 \times 10^{-19}/1$

Electricity MCQ (class X)

n = (4/1.6)×10¹⁹

n = 2.5×10¹⁹

46) (b). I = nq/time

2 = n ×1.6×10⁻¹⁹/16

n = 2×10¹⁸

47) (c). reciprocal of conductance if resistance and its unit is moh (Siemens/m)

48) (b). Power dissipation is a dissipation of heat which is proportional to resistance. Due to increase in length 100% the resistance will become double so heat dissipation becomes double. So the answer is 200%.

49) (b). Power = Voltage × Current

800 = 200 × I

I = 4 Amp

R = V/I = 200 / 4 = 50 Ω

50) (c). Equivalent resistance of the series connection: 2+4 = 6Ω

Voltage = 6V

I = 6/6 = 1Amp.

H = I²RT = 1²×4×5 = 20J

51) (a). Van De Graff generator could develop 200000 volt of static electricity.

52) (a). Current is proportional to voltage.

53) (c). High voltage is used to reduce current and heat loss due to the transfer of electricity.

54) (a). H = I²RT See also Q-36

55) (c). Amp = total charge / time in sec.

1 = Q/10×60

Q = 600 coulomb.

56) (a). resistance is proportional to length.

Let the length of Nicrome wire be L.

So. 5/L = 2.5/120

L = 600/2.5 = 240 Cm.

57) (d).

58) (b). Theoretically it should be zero.

59) (a). Due to discontinuity in the circuit the resistance should be infinity.

60) (c). The resistance of a resistor is inversely proportional to the area of cross section of the resistor.

61) (b). Semiconductors have a negative temperature coefficient of resistance. Due to increase in temperature number of free electrons are increased which are the cause behind the increase in its current.

62) (a). Due to increase in resistance the conductance decreases.

63) (c).

64) (b). Semi and super conductors do not follow Ohm's law.

65) (a). Equivalent resistance when connected in parallel = E_p

$1/E_p = 1/5 + 1/5 + 1/5$

$E_p = 5/3$ Ω

Equivalent resistance when connected in series = E_s

$E_s = 5+5+5 = 15Ω$

Now $E_p/E_s = (5/3)/15 = 1/9$ Ω

66) (b). Street lights are connected in parallel so that they all could bet same voltage and failure of any one street light should not affect others.

67) (c). The resistance of supply line is very less compared to heater coil and heater develop in also very negligible therefore.

68) (a). According to Lenz law its happen due to resistance.

69) (c). Two parallel resistance equivalent resistance become

$1/R_p = 1/2 + 1/2$ $R_p = 1\Omega$

When its connected with other resistance in series total circuit resistance become
$R_e = 1+2 = 3\Omega$

70) (a).

71) (b). Here Graphite is the only conductor. Dry HCl (hydrogen chloride gas) & Carbon black does not carry electricity as it does not have free charges. Rubber is an insulator.

72) (d). Ammeter is connected in parallel o the circuit and heat dissipation is less during parallel connection due to reduction of resistance of the circuit.

73) (a). Ohmmeter measure the resistance. When it indicates zero reading that means the wires are connected or shorted. Open circuit resistance will be infinity.

74) (c). Area of the cross section is inversely proportional to the resistance.

75) (a). Solar cells are made of silicon which is a semiconductor material.

76) (d). Nothing is called semi insulator in physics.

77) (c). Fuse wire should melt instantaneously so that fault current should not pass through the circuit.

78) (a). Resistance is proportional to the length of the wire.

79) (b). $P = VI$

$P = V \cdot V/R$

$P = V^2/R$

When V is constant Power is inversely proportional to Resistance. So 100W bulbs will have more resistance.

80) (a). The ideal resistance of a voltmeter should be infinity so that no current should flow through it and voltage will not drop.

81) (a).

82) (b). Two terminals. Due to two phase connection.

83) (b). Three terminals. Due to three phase connection.

Magnetic Effect of Electric Current

1) **Electric current passing through the copper wire will produce**

   a) Magnetic field.
   b) Electric field.
   c) Both (a) and (b).
   d) None of the above.

2) **Magnetic field is**

   a) Vector quantity.
   b) Scalar quantity.
   c) A physical quantity.
   d) Imaginary quantity.

3) **Magnetic field lines are generated from**

   a) South pole.
   b) North pole.
   c) Either north of south pole depends on the material of the magnet.
   d) Always start from the centre of the magnet.

4) **The magnetic force will be strongest when the direction of current is _____ to the direction of magnetic field.**

   a) 0°
   b) 45°
   c) 90°
   d) 180°.

5) **Between electro magnet and a permanent magnet which one is more power full for the same size**

   a) Electromagnet.
   b) Permanent magnet.
   c) Both are same.
   d) Cannot be compared.

6) **The split ring is a standard component of**

   a) Electric generator.
   b) Electric motor.
   c) Both (a) and (b).
   d) None of the above.

7) **Presence of magnetic field can be determined by**

   a) Voltmeter.
   b) Rheostat.
   c) Galvanometer.
   d) Ammeter.

8) **Which of the following works on Flemings left hand rule**

   a) Loud speaker.

b) Motor.
c) Generator.
d) Both (a) and (b)

9) Who first observed the magnetic effect of electric current

a) Orested.
b) Faradey.
c) Ampere.
d) Gauss.

10) Choose the incorrect statement(s) regarding magnetic lines of the field

a) The direction of magnetic field at a point is taken to be the direction in which the north pole of a magnetic compass needle points.
b) Magnetic field lines are closed curves.
c) If magnetic field lines are parallel and equidistant, they represent zero field strength.
d) The relative strength of magnetic field is shown by the degree of closeness of field lines.

11) The phenomena of electromagnetic induction is

a) The process of charging a body.
b) The process of generating magnetic field due to a current passing through the coil.
c) Producing induced current in a coil due to relative motion between a magnet and the coil.
d) The process of rotating a coil of an electric motor.

12) The core of an electromagnet should be made of

a) Soft iron.
b) Hard iron.
c) Either soft or hard iron.
d) Alnico.

13) When current flows in the Anti-clockwise direction in a current carrying conductor, the pole it creates is

a) South pole.
b) North pole.
c) Neutral point.
d) None of the above.

14) In the experiment of electromagnetic induction, the ____coil is attached to the galvanometer

a) Primary coil.
b) Secondary coil.
c) Tertiary coil.
d) Either Primary or secondary coil, depends on the arrangement.

15) The relative field strength of magnetic field is compared by

a) Flux.
b) Lumen.
c) Candela.
d) Farad.

16) The magnetic field produced by a current carrying straight wire depends

a) Inversely on the distance from it.
b) Proportionately on the distance from it.
c) Inversely on the square of the distance from it.
d) Proportionately on the square distance from it.

17) When current flows in the clockwise direction in a current carrying conductor, the pole it creates is

a) South pole.
b) North pole.
c) Neutral point.
d) None of the above.

18) For a current in a long straight solenoid N and S poles are created at the two ends. Among the following statements, which one is incorrect?

a) The field lines inside the solenoid are in the form of straight lines, which indicates that the magnetic field is the same at all points inside the solenoid.
b) The strong magnetic field produced inside the solenoid can be used to magnetize a piece of magnetic material such as soft iron, when placed inside the coil.
c) The pattern of the magnetic field associated with the solenoid is different from the pattern of the magnetic field around a bar magnet.
d) The N and S pole exchange position when the direction of current through the solenoid is reversed.

19) **The acceptable frequency of AC in India**

a) 50Hz.
b) 60Hz.
c) 40Hz.
d) Either 50Hz or 60Hz.

20) **The essential difference between a Dc generator and an AC generator is**

a) Dc generator has permanent magnet where as AC generator has an electromagnet.
b) DC generators have commutator but AC generator has slip rings.
c) AC generator will generate higher voltage.
d) DC generator will generate higher voltage.

21) **In all the electrical appliances, the switches are put in the**

a) Earthen wire.
b) Live wire.
c) Neutral wire.
d) Any of the above.

22) **The magnetic field produced in a straight conductor on passing current through it is**

a) In the direction of the current.
b) In the opposite direction of the current.
c) Parallel to the wire.
d) Circular around the wire.

23) **Maxwell's Cork screw rule is also called**

a) Right hand thumb rule.
b) Left hand thumb rule.
c) Right hand rule.
d) None of the above.

24) **What is the shape of field lines of a magnetic field passing through the centre of current carrying circular ring?**

a) Circular.
b) Ellipse.
c) Straight line.
d) Parabolic.

25) **The device used for producing electric current is**

a) A dynamo.
b) A motor.
c) Ammeter.
d) Rheostat.

26) **Commercial electric motor do not use**

a) An electromagnet to rotate the armature.
b) Effectively large number of turns of conducting wire in the current carrying coil.
c) A permanent magnet to rotate the armature.
d) A soft iron core on which the coil is wound.

27) **On which electromagnet strength of a solenoid is not depends**

a) Amount of current passing through the coil.
b) Number of turns of the coil.
c) Type of core used.
d) Operation time.

28) **A magnetic field exerts a _____ on a current carrying conductor.**

a) Electromotive force.
b) Magnetic force.
c) Induction current.
d) None of the above.

29) Magnetic lines are indicating

a) The direction of the magnetic field.
b) The direction of current due to which field was developed.
c) Opposite direction of magnetic field.
d) None of the above.

30) Armature made of soft iron due to

1) Easily could magnetize and demagnetized by switching On and Off.
2) Can reverse the pole easily.
3) Power level could change easily.
4) Could make less powerful than the permanent magnet.

a) 1, 2 & 4 are correct.
b) 2, 3 & 4 are correct.
c) 1, 2 & 3 are correct.
d) All are correct.

31) Choose the incorrect statement

a) Fleming left hand rule is a simple rule to know the direction of induced current.
b) He right hand thumb rule is used to finding the direction of magnetic fields due to current carrying conductor.
c) The difference between the DC and AC is that the DC always flows in one direction where as the AC is reversed the direction periodically.
d) In India AC change direction after every 1/50 sec.

32) To transfer electric power to a large distance which types of current are most efficient?

a) AC
b) DC
c) Both AC and DC are equivalent in terms of energy loss.
d) Could not be compared.

33) At the time of short circuiting the current in the circuit

a) Reduces substantially.
b) Increase heavily.
c) Vary randomly.
d) Remains constant.

34) The solar cell is a source of

a) AC.
b) DC.
c) Either Ac or DC.
d) Its not a source of current.

35) In a galvanometer zero is at

a) The left end of the dial.
b) The right end of the dial.
c) Middle of the dial.
d) Either left of the right end of the dial.

36) If two free parallel conductors are carrying current in a different direction and located near to each other then

a) They will attract each other.
b) They will repeal each other.
c) They will not act any force to each other.
d) Get moved to perpendicular to each other.

37) MRI of the brain done for some dieses. Here MRI stands for

a) Magnetic Resonance Imaging.
b) Maximum Resonance Imaging.
c) Minimum Resonance Imaging.
d) Metallic Resonance Imaging.

38) Which one is a false statement

a) An electric dynamo converts mechanical energy o electrical energy.
b) An electric generator works on the principle of electromagnetic induction.
c) The field at the centre of the long circular coil carrying current will be perpendicular to straight lines.

d) A circular current carrying conductor affected by a magnetic field.

39) In an electric motor a _____ act as a commutator

a) Slip ring.
b) Split ring.
c) Solenoid.
d) Splitter.

40) A rectangular coil of copper wires is rotated in a magnetic field. The direction of induced current changes once on each

a) Half revolution.
b) One revolution.
c) Two revolutions.
d) Never change direction.

41) Neutral wire color code is

a) Blue.
b) Black.
c) Green.
d) Red.

42) The direction of magnetic field lines in the region outside the bar magnet is

a) From N pole towards the S pole of the magnet.
b) From S pole towards the N pole of the magnet.
c) In the direction coming out from both the poles of the magnet.
d) In the direction entering poles in the magnet.

43) According to Flemings Left hand rule the thumb shows

a) The direction of the magnetic field.
b) The direction of current.
c) The direction of force.
d) The Direction of induced current.

44) According to Flemings Left hand rule the middle finger shows

a) The direction of the magnetic field.
b) The direction of current.
c) The direction of force.
d) The direction of induced current.

45) According to Flemings Left hand rule the index finger shows

a) The direction of the magnetic field.
b) The direction of current.
c) The direction of force.
d) The direction of induced current.

46) Which is/are a source of direct current

a) Solar cell.
b) Daniel cell.
c) Dynamo.
d) All of the above.

47) The strength of magnetic field inside a long current carrying straight solenoid is

a) More at the ends than at the center.
b) Minimum in the middle.
c) It is same at all points.
d) Found to increase from one end to another.

48) The direction of flow of current in a circuit could be easily reversed by using

a) Commutator.
b) Rheostat.
c) Galvanometer.
d) Chronometer.

49) The magnetic field of a bar magnet is comparable with

a) A current carrying wire.
b) A current carrying ring.
c) Current carrying solenoid.
d) Current carrying rectangular loop.

50) The magnetic field produced in a circular coil when the current

passing through it is inversely proportional to

a) Length of the coil.
b) The radius of the coil.
c) The weight of the coil.
d) The current of the coil.

51) **To convert an AC generator into a DC generator**
a) Split ring type commutator must be used.
b) Slip rings and brushes must be used.
c) A stronger magnetic field has to be used.
d) A rectangular wire loop has to be used.

52) **1 gauss =**
a) 10 Tesla.
b) 100 Tesla.
c) 10^{-4} Tesla.
d) 10^{4} Tesla.

53) **A current through a horizontal power line flows in the north to south direction, then the direction of magnetic field at the point just directly below it will be**
a) West to east.
b) East to the west.
c) South to the north.
d) North to south.

54) **A current through a horizontal power line flows in the north to south direction, then the direction of magnetic field at the point just directly above it will be**
a) West to east.
b) East to the west.
c) South to the north.
d) North to south.

55) **The magnetic field produced at the Centre of a circular coil does not depend on**
a) The radius of the loop.
b) Current magnitude.
c) The weight of the coil.
d) Both (a) and (b)

56) **As we move away from the current carrying conductor the spacing between the magnetic lines is**
a) Increases.
b) Decreases.
c) Remains constant.
d) May decrease and it depends on the geographical magnetic field.

57) **If the number of turn in a solenoid is doubled then magnetic field strength will**
a) Doubled.
b) Halved.
c) Quadrupled.
d) One forth.

58) **Magnetic field has**
a) Direction only.
b) Magnitude only.
c) Both magnitude and direction.
d) Neither direction nor magnitude.

59) **Two poles of a magnet have**
a) Equal strength.
b) Unequal strength.
c) Can have same or different strength.
d) None of these.

60) **Magnetic field lines are**
1) Form closed curve.
2) Towards pole field lines are more per unit area.
3) Field lines could intersect each other at the centre of the magnet.
4) Field lines are generated from the South Pole.

a) 1, 2 and 3 are correct.
b) 1 and 4 are correct.
c) 1, 2 and 4 are correct.
d) 1 and 2 are correct.

61) **The most important safety method used for protecting home appliances from short circuiting or overloading is**

a) Earthing.
b) Use of fuse.
c) Use of stabilizer.
d) Use o electric motor.

[Answers & Explanations]

1) **(a).** When electric current passing through the copper wire it will produce a magnetic field.

2) **(a).** The magnetic field is having both direction and magnitude so it is a vector quantity.

3) **(b).** Magnetic field lines are started from north pole and go to the south pole.

4) **(c).** Magnetic field force F=qvB Sinθ q is a charge, v= velocity and B is magnetic field strength. Sinθ is maximum when θ=90°.

5) **(a).** The strength of an electromagnet is more compared to the permanent magnet of the same size.

6) **(b).** The split ring in the electric motor also known as a commutator reverses the direction of current flowing through the coil after every half of coil rotation.

7) **(c).** Due to the flow of current magnetic field is produced and that deflects the galvanometer.

8) **(d).** Generator follows Flemings right Hand rule.

9) **(a).**

10) **(c).** If the magnetic field lines are parallel and equidistant, they represent equal field strength , not zero field strength. All other statements are correct.

11) **(c).**

12) **(a).** The core of electromagnet made of soft iron so that it could magnetize and demagnetized as per requirement. If we use hard iron then it will become a permanent magnet.

13) **(a).** As per Maxwell cork screw law.

14) **(b).**

15) **(a).** Flux is magnetic field lines passing through the area. Its SI unit is Weber and CGS unit is Maxwell.

16) **(c).** According to Bio-Savart's rule the magnetic field produced by a current carrying straight wire is inversely proportional to the square of the distance from it.

17) **(b).** According to the Maxwell Cork screw rule.

18) **(c).** The pattern of the magnetic field associated with the solenoid is same as the pattern of the magnetic field around a bar magnet.

19) **(a).**

20) **(b).** A commutator is a rotary electrical switch in certain types of electric motors and electrical generators that periodically reverses the current direction between the rotor and the external circuit

21) **(c).**

22) **(d).**

23) **(a).** Maxwell's cork screw rule states that if a screw is placed in the direction of current in then the rotation sense corresponding to the current will give the magnetic field.

24) **(c).**

25) **(a).**

26) (c).

27) (d). Solenoid strength depends on all given options except Operation time.

28) (b).

29) (a).

30) (c). Electromagnets are more power full than the permanent magnet.

31) (a). Fleming Right hand rule is used to know the direction of induced current.

32) (a). Because high voltage could be produced by AC and that reduces the loss through the conductor as at high voltage current become less. Due to less current flow Joule heat loss will be low.

33) (b). During short circuit the resistance becomes very very less and theoretically it becomes zero.

34) (b).

35) (c). Zero is marked at the middle so that needle could move in both the direction according to the flow direction of current through the conductor.

36) (b). When two parallel conductors are carrying current in a different direction they will repeal each other.

37) (a).

38) (c). It is parallel.

39) (b).

40) (a). That is why it produces alternating current.

41) (b). According to international color code Neutral wire color is Black.

42) (a).

43) (c). According to Flemings Left hand rule if thumb, middle finger and forefinger are stretched mutually perpendicular to each other then if forefinger finger shows the direction of magnetic field, middle finger shows the direction of current then thumb should indicate the direction of motion or the force acting on the conductor.

44) (b). See Q-43.

45) (a). See Q-43

46) (d).

47) (c).

48) (a).

49) (c).

50) (b).

51) (a).

52) (c).

53) (a). According to crock screw rule.

54) (b). According to crock screw rule.

55) (c). Magnetic field strength does not depends on the weight of the coil.

56) (a). Space between the magnetic lines increases as magnetic field strength decreases.

57) (a). Magnetic field strength is proportional to the number of turns.

58) (c). The magnetic field is a vector quantity.

59) (a).

60) (d). Magnetic field lines never crossed each other as at same point direction of magnetic field should not be twice. Field lines are generated from the north pole.

61) (b) Fuse wire disconnects the fault circuit from the source.

SOURCE OF ENERGY

1) Which of the following is a major source of energy

a) Coal.
b) Petroleum.
c) Nuclear.
d) Wind.

2) The ultimate source of energy to human is

a) Water.
b) Sun.
c) Wind.
d) Nuclear.

3) The nuclear fuel used in nuclear reactors are

a) Uranium and Plutonium.
b) Thorium and Plutonium.
c) Uranium and Thorium.
d) Uranium and Thorium.

4) Which source of energy is different from other

a) Wind.
b) Coal.
c) Petroleum.
d) Biogas.

5) The coolant used in Nuclear power in India is

a) Heavy water only.
b) Light water only.
c) Light Water or Heavy water.
d) Sodium metal.

6) Which of the following power is not responsible for acid rain

a) Coal.
b) Petroleum.
c) Natural gas.
d) Solar power.

7) Source of energy of the sun is

a) Nuclear fission.
b) Nuclear fusion.
c) Chemical reaction.
d) All of the above.

8) The radiation emits by the sun is responsible for skin cancer is

a) Micro waves.
b) X-rays.
c) Gama rays.
d) UV rays.

9) Choose incorrect statement(s)

1) Gobar Gas is produced when crops, vegetable wastes decompose in presence of oxygen.
2) Main gas contains in Bio-gas is ethane.
3) Bio-mass is a renewable source of energy.

a) 1 & 2 are incorrect.
b) 2 & 3 are incorrect.
c) 1 & 3 are incorrect.

d) All are incorrect.

10) _____ is the main part of the solar cooker to generate heat.

a) Glass sheet.
b) Black color coating inside the box.
c) Mirror.
d) None of the above.

11) Which of the following statement(s) is correct regarding wind power

1) In rainy season power generation is more due to more air mass hitting the blades.
2) Power generation depends on air velocity.
3) Power generation id proportional to height.

a) All are correct.
b) 1 and 2 are correct.
c) 2 and 3 are correct.
d) Only 2 is correct.

12) The potentiometer is used to measure

a) Potential of the cell.
b) The voltage of the cell.
c) The resistance of the cell.
d) The conductance of cell.

13) Hot furnace radiate

a) UV rays.
b) X-rays.
c) Gama rays.
d) Infra red rays.

14) A major problem in generation of nuclear power is

a) Fission of nucleolus.
b) Unstable nuclear reaction.
c) Dispose of nuclear wastages.
d) Mining of nuclear fuel.

15) Many thermal power plants are located near the coal and oil fields because

1) Transmission of electricity compare to petroleum and coal is easy and efficient.
2) Fuel transmission cost is less.
3) Produce less pollution.

a) 1 & 2 are correct.
b) 2 & 3 are correct.
c) 1 & 3 are correct.
d) All are correct.

16) 1KWh =

a) 4.2×10^6 Joule
b) 3.2×10^6 Joule
c) 3.6×10^6 Joule
d) 1.8×10^6 Joule

17) Following is the main source of geothermal energy

a) Lava.
b) Magma.
c) Geyser.
d) All of the above.

18) The sluice gate is used in

a) Nuclear power plant.
b) Gas based power plant.
c) Oil based power plant.
d) Hydropower plant.

19) Acid rain occurs due to use of

a) Nuclear power.
b) Coal based thermal power.
c) Solar power.
d) Hydro power.

20) Large eco-system will be submerged if we try to develop

a) Dam for the hydro power plant.
b) Gas based thermal power plant.
c) Oil based thermal power plant.
d) Wind power plant.

21) Total number of a commercially operating Nuclear power plant in India (at present)

a) 15.
b) 16
c) 18
d) 21

22) Tehri dam is located on

a) Mahanadi.
b) The Godavari.
c) Ganges river.
d) Narmada river.

23) The temperature difference between Ocean layers is required to install OTEC power plant is

a) 20°C.
b) 30°C.
c) 40°C.
d) 50°C.

24) Which of the following is not a conventional source of energy

a) Wind energy.
b) Bio gas energy.
c) Nuclear.
d) Coal.

25) In hydro power plant production of electricity will follow the following the path

a) Potential energy-linear Kinetic energy-Rotational kinetic energy-Electrical energy.
b) linear Kinetic energy- Potential energy-Rotational kinetic energy-Electrical energy.
c) Rotational kinetic energy-Potential energy-linear Kinetic energy- Electrical energy.
d) linear Kinetic energy-Rotational kinetic energy- Potential energy-Electrical energy.

26) The floating generator is used to generate

a) Tidal energy.
b) Ocean energy.
c) Wave energy.
d) Hydro power.

27) Sardar Sarovar project is located on___ river.

a) Mahanadi.
b) The Narmada.
c) Tapti.
d) Kaveri.

28) Choose the correct statement(s)

1) Monazite is a source of Uranium.
2) Gobar gas mainly contains methane.
3) Maximum power generated in India from coal.
4) DVC is famous for Hydro power plant.

a) 1,2 & 3 are correct.
b) 2 ,3 & 4 are correct.
c) 1 & 3are correct.
d) All are correct.

29) On which river Bhakra Nangal dam is situated

a) Ganges river.
b) Beas river.
c) Jhelum.
d) Ravi river.

30) Ocean thermal energy is due to

a) Energy in waves.
b) The temperature gradient in a different layer of the ocean.
c) Pressure gradient in different layer of ocean.
d) All of the above.

31) The lid of solar cooker should make of

a) Plastic.
b) Bakelite.
c) Glass.
d) Metal.

32) When wood burns in a limited supply of oxygen it produces

a) Active charcoal.
b) Charcoal.
c) Coal.
d) Carbon soot

33) Solar cells are made of

a) Copper.
b) Silicon.
c) Carbon.
d) Nickel.

34) India's largest wind power production unit is located at

a) Kudankulam.
b) Surat.
c) Darjeeling.
d) Mangalore.

35) Solar cooker use

a) Convex mirror.
b) Concave mirror.
c) Plane mirror.
d) Plano concave mirror.

36) Charcoal burn_____ flame and comparatively_____, compare to coal.

a) With, more smoke.
b) Without, more smoke.
c) With, smokeless.
d) Without, smokeless.

37) Which of the following is the cleanest source of energy

a) Nuclear.
b) Wind.
c) Solar.
d) Water.

38) Which of the following is not renewable source of energy

a) Nuclear.
b) Solar.
c) Hydro.
d) Bio-gas.

39) Which one is not created greenhouse effect

a) CO_2.
b) SO_2.
c) CH_4.
d) H_2O

40) Coal is
a) 2 types.
b) 3 types.
c) 4 types.
d) 5 types.

41) The threshold air velocity for wind energy production is

a) 10 km/hr.
b) 15 km/hr.
c) 05 km/hr.
d) 20 km/hr.

42) The fuels formed from decomposed organic materials under the earth surface is called

a) Bio gas.
b) Natural gas.
c) Fossil fuel.
d) Bio mass.

43) Which type of coal is mostly use in coal based thermal power plant

a) Anthracite.
b) Bituminous.
c) Lignite.
d) Pit.

44) A box type solar cooker could attain maximum temperature

a) Up to 60°C.
b) up to 100°C.
c) up to 150°C.
d) up to 250°C.

45) Where is nuclear power plant not located?

a) Madras(Chennai)
b) Kalpakkam.
c) Rawatbhatta.
d) Rana pratap sagar.

46) Which energy generated on earth is not derived from the sun

a) Hydel power.
b) Wind power.
c) Nuclear power.
d) Bio gas power.

47) In solar power panel which of the following metal is used

a) Gold.
b) Titanium.
c) Silver.
d) Nickel.

48) Which conventional source of energy fully depends on greenhouse gas

a) Thermal.
b) Bio gas.
c) Nuclear.
d) Hydro power.

49) One solar cell could develop a voltage of

a) 0.5 to 1V.
b) 1 to 1.5V.
c) 1to 2V.
d) 0.1 to 0.6V.

50) Slurry left after producing "Gobar gas" is used as

a) Manure.
b) Domestic fire source.
c) Medicine.
d) Chemical for the cement industry.

51) Manure is rich of

a) K and P.
b) N_2 and K.
c) N_2 and Ca.
d) N_2 and P.

52) Bio gas mostly contains

a) Ethane.
b) Methane.
c) Butane.
d) CO_2 gas.

53) Both power and manure produced by

a) Bio gas plant.
b) Thermal power plant.
c) Nuclear power plant.
d) Hydel power plan.

54) Bio gas contains methane up to

a) 65%.
b) 75%.
c) 85%.
d) 95%.

55) Each solar cell could produce energy up to

a) 1.5w.
b) 2w.
c) 0.5w.
d) 0.75w.

56) Solar constant is

a) Solar energy reaching on earth surface per unit area, perpendicular to rays at the average distance between sun and earth.
b) It is the average distance between sun and earth.
c) It is the average energy reaching on earth per year.
d) It is the total energy reaching on the earth per year.

57) Value of solar constant is

a) 1.4 w/m²
b) 1.4 Kw/m²
c) 1.4 w/cm²
d) 14 w/m²

58) Which power producing depends on earth gravity

a) Wind power.
b) Solar power.
c) Gas power.
d) Hydel power.

59) Which power producing unit required the lowest maintenance

a) Wind power.
b) Solar power.
c) Gas power.
d) Hydel power.

60) 1 amu is equivalent to

a) 93.1 MeV
b) 0.931 MeV.
c) 931MeV.
d) 1000MeV.

[A COMPLETE GUIDE TO MCQ]

CHAPTER: 14

[Answers & Explanations]

1) **(a)**. Coal is the major source of energy.

2) **(b)**. Sun is the ultimate source of energy for human beings. Except for Nuclear energy all came from Sun.

3) **(d)**. Uranium and Thorium are the main nuclear fuels at present time.

4) **(b)**. Except for wind all others are fossil fuel.

5) **(c)**. Indian reactors are using either light water (H_2O) of heavy water (D_2O) as a coolant.

6) **(d)**. Solar power does not produce any gases which are responsible for acid rain.

7) **(b)**. Due to the fusion of H to He solar energy is produced.

8) **(d)**. UV rays are the mainly responsible skin cancer.

9) **(d)**. Gobar gas is produced from gobar. The bio gas main constituent is Methane and Bio-mass is not a renewable source of energy but it is a non-conventional source of energy.

10) **(a)**. Glass sheet reflects back low energy heat waves due to larger wave length.

11) **(d)**. Wind turbine develops power at certain speed and to maintain that speed wind velocity has to maintain higher than 15 km/Hr.

12) **(a)**. To measure the potential difference of cell.

13) **(d)**. Hot furnace never radiate first three rays except for infra-red rays.

14) **(c)**. Dispose of nuclear waste is the main problem in generation of nuclear power.

15) **(a)**. Pollution is not depended on the location of the thermal power plant.

16) **(a)**.

17) **(b)**. Hot magma is the main source of heat for hot spring.

18) **(d)**. Sluice get supply water in controlled rate to hydel power turbine.

19) **(b)**. The coal based power plant produces CO2, NO2, SO2 gas which are the main cause of acid rain. Another source of energy does not produce such gases.

20) **(a)**. Construction of dam for the hydro power plant is submerged large local ecosystem.

21) **(c)**. 6 in Rawatbhatta (Rajasthan). 4 in Tarapur (Maharashtra), 4 in Kaiga (Karnataka), 4 in Madras and Kudankulam (Tamilnadu). 2 in Kakrapara (Gujrat) now under re-construction.

22) **(c)**. Tehri dam is located in Uttarakhand on Bhagirathi river.

23) **(a)**. To install OETC power plant ocean layer temperature should be maintained 20°C.

24) **(c)**. Nuclear energy is the non-conventional source of energy.

25) **(b)**. **Conversion of energy:** potential energy of water stored in Dam to the kinetic energy of water flow to the rotational kinetic energy of turbine to Electric energy in the generator.

26) **(c)**. Wave energy stored by floating generator.

[A COMPLETE GUIDE TO MCQ]
CHAPTER: 14

27) (b). It located in Gujarat and dam constructed on Narmada river.

28) (b). Monazite is the main source of thorium. Uraninite (earlier named as Pitch blend) is the name of Uranium ore.

29) (b). On Beas river in Punjab.

30) (b).

31) (c). Please refer Q-10

32) (b). Activated charcoal is produced from coconut shell burning in a limited supply of oxygen. Coal formed by dead giant plant layer in presence of heat and pressure. Carbon soot produced by fossil fuel burning.

33) (b). Silicon is a semi-conductor and its use to make solar cell doped with gallium.

34) (a). Kudankulam is located in Tamilnadu.

35) (c). To trap the long heat waves by reflecting it back to the cooker.

36) (d). Charcoal burns without flame as it does not contain any gaseous substances in it to burn.

37) (b). Wind energy does not affect the environment but it is hazardous to birds. Nuclear energy produces nuclear waste and radioactive effluents. Water power submerges large ecosystem. Solar power pollute the atmosphere during the preparation of solar cells.

38) (d). All other sources are renewable except Bio-gas.

39) (b). Sulpher di-oxide gas produces acid rain but not a green house gas.

40) (c). Coal is of 4 types. Anthracite, Bituminous, Lignite and peat.

41) (b). When wind velocity is less than 15 Km/hr, a wind turbine could not achieve critical speed so it is the threshold value of air velocity for wind power generation.

42) (c). Fossil fuels are formed from decomposed organic materials (giant plant, animals etc.) under the earth surface with pressure and heat.

43) (b). Anthracite is used for iron extraction. Lignite and peat mostly use for domestic fuel. Bituminous coal has the specific quality to use as thermal power plant fuel.

44) (c). Modern solar cooker cell could attain maximum 150° C.

45) (d). Rana pratap sagar is famous for Hydro power plant.

46) (c). Nuclear power is not related to energy comes from the sun.

47) (c). The silver used in the solar cell works as a conductor to collect these electrons in order to form a useful electric current.

48) (b). The main constituent of Bio-gas is Methane gas and it is a greenhouse gas.

49) (a). The modern solar cell could develop 0.5 to 1Volt.

50) (a).

51) (d).

52) (b).

53) (a). Bio gas plant produces methane as a source of power and remaining slurry is used as manure.

54) (b).

55) (d).

56) (a). Solar energy reaching on earth surface per unit area, perpendicular to rays at the average distance between Sun and Earth is called solar constant.

57) (b).

58) (d). Potential energy is the product of mass, Gravity and height of the mass from the earth surface. This potential energy of huge water is used to produces hydro power.

59) (b). Solar power plant requires minimal maintenance due to lack of moving parts compared to other power plant equipment's.

60) (c).

Our Environment

CHAPTER: 15

1) **In producers energy is stored as**
   a) Kinetic energy.
   b) Heat energy.
   c) Chemical energy.
   d) Potential energy.

2) **In ecosystem nektons are**
   a) Small floating animals in the ecosystem.
   b) Animals that can swim and navigates in the ecosystem.
   c) Algae, fern etc.
   d) All producers.

3) **The functional unit of the environment is**
   a) Ecosystem.
   b) Biome.
   c) Ecological niche.
   d) Biomass.

4) **In an ecosystem the biomass of abiotic components compared to biotic components biomass is.**
   a) equal.
   b) Less.
   c) More.
   d) Slightly less.

5) **Each step of the food chain is called**
   a) Trophic level.
   b) Lindeman Energy level.
   c) Connectors.
   d) Nekton

6) **To protect the whole ecosystem we should try to protect mostly**
   a) Apex consumer.
   b) Intermediate consumer.
   c) Producers.
   d) Decomposers.

7) **Which of the following is an example of an artificial ecosystem**
   1) Aquarium.
   2) Crop field.
   3) Lake.
   4) Pond.

   a) 1 & 2 are incorrect.
   b) 2 & 3 are incorrect.
   c) 1 & 3 are incorrect.
   d) All are incorrect.

8) **Which of the following is Unidirectional in the ecosystem**
   a) Material flow.
   b) Energy flow.
   c) Food chain.
   d) All of the above.

9) **Which garbage disposal method produces manure?**
   a) Composting.
   b) Land filling.
   c) Open dumping.
   d) Recycling.

10) **In Aquatic ecosystem primary carnivorous are**

a) Phytoplankton.
b) Zooplankton.
c) Small fish.
d) Large fish.

11) In a food chain the third level is always occupied by

a) Decomposers.
b) Produces.
c) Herbivorous.
d) Carnivorous.

12) Which of the following statement(s) is correct regarding Ozone layer

1) It protects us from harmful UV rays
2) Ozone gas is allotrope of Oxygen gas.
3) The ozone layer is broken by UV rays.
4) The ozone layer is depleted due to CFC.

a) All are correct.
b) 1,2 and 3 are correct.
c) 1,2 and 4 are correct.
d) 2 and 3 are correct.

13) Incineration is a method of

a) Food production method by blue green algae.
b) Energy transformation method in an ecosystem.
c) Waste disposal.
d) Waste control.

14) Detritus food chain is

a) Start from the dead organic matter to the detrivore organisms, which in turn make food for protozoan to carnivores etc.
b) Starts from the green plants that make food for herbivores and herbivores in turn for the carnivores.
c) Take care of the dead remains of organisms at each trophic level.
d) None of the above.

15) Ozone is not the cause of

a) Diarrhea
b) Cataract.
c) Crop failure
d) Skin cancer.

16) _____ is not a biodegradable pollutant

a) Cotton cloth.
b) Washing powder.
c) Paper mold.
d) Stool.

17) Which of the following statement(s) is/are correct

1) Food web is never straight.
2) Food web defines only one source of food.
3) Complex food web brings better stability.
4) Each step of food web indicates trophic level.

a) 1,2 & 3 are correct.
b) 2 & 4 are correct.
c) 1,3 & 4 are correct.
d) Only 2 is correct.

18) Green plants could convert ____% energy to food energy

a) 1
b) 10
c) 1.5
d) 2

19) Which one is not unidirectional

a) Food web.
b) Food chain.
c) Energy in the ecosystem.
d) None of the above.

20) The flow of material in the ecosystem is

a) Unidirectional.
b) Multidirectional.
c) Bi directional.
d) Cyclic.

21) Food chains are

a) Only one types.
b) Two types.
c) Three types.
d) Four types.

22) Which protocol control production and use of CFCs in the world.

a) Kyoto protocol.
b) Vienna protocol.
c) Copenhagen protocol.
d) Montreal protocol.

23) Ozone layer depleted due to

a) Coal Industries.
b) Automobiles.
c) Forest fire.
d) Refrigerant industries.

24) Number of decomposers compared to producers is

a) Many times more.
b) Slightly more.
c) Approximately same.
d) Very very less.

25) An eagle consumed a snake and got 5 kJ energy. What will be the energy produced in producer if the chain follows below

   Producer – Grass hopper-Frog – Snake-Eagle

a) 50 KJ
b) 500 KJ
c) 5000 KJ
d) 50000 KJ

26) The inverted ecological pyramid is provided by

a) Parasitic food chain.
b) Saprophytic food chain.
c) Energy pyramid.
d) Pyramid of biomass.

27) The main function of decomposers in our ecosystem is

a) To convert inorganic materials to simple form.
b) To convert inorganic materials to organic materials.
c) To convert organic materials into inorganic materials.
d) To breakdown organic complex compounds.

28) An ecosystem includes

a) All living beings only.
b) All non living beings only.
c) All living and non livings.
d) Some living organisms and some non living organisms which are directly interlinked.

29) If from one trophic level to another next trophic level only 10% energy transferred then remaining 90% energy

a) Lost in the ecosystem.
b) Use by the preset trophic level organisms for their life process.
c) Converted to heat energy.
d) Indirectly transferred to the sun.

30) Which group of organisms are not constituents of a food chain?

1) Grass, lion, rabbit, wolf.
2) plankton, man, fish, grass hopper.
3) Wolf, grass, snake, tiger.
4) Frog, snake, eagle, grass, grass hopper.

a) 1 & 2 are correct.
b) 3 & 4 are correct.
c) 2 & 3 are correct.
d) None of the above.

31) Grasshopper-Frog–Snake-Eagle in this food chain if frog suddenly disappeared then

a) The population of Eagle will decrease but snake will increase.

b) The population of Eagle will decrease but Grasshopper will increase.
c) Eagle will start eating grass hoppers.
d) Food chain will affect all the living organisms directly or indirectly.

32) CFCs are harmful to

a) Ozone layer.
b) Exosphere.
c) Troposphere.
d) Magnetosphere.

33) In an ecosystem _____ % of energy could transfer from one tropic level to next tropic level

a) 5
b) 10
c) 15
d) 20

34) CFC stands for

a) Carbon flouro carbon.
b) Calcium chloro carbon.
c) Chloro Fluoro Carbons.
d) Carbon flouro chloro.

35) Food chains mainly have

a) 2 to 3 trophic levels.
b) 3 to 4 trophic levels.
c) 4 to 5 trophic levels.
d) 5 to 6 trophic levels.

36) In ecosystem flow of energy is always

a) Unidirectional.
b) Bi-directional.
c) Multidirectional.
d) Depends on the trophic level.

37) Following element ion is mainly responsible for Ozone layer depletion

a) Fluorine.
b) Chlorine.
c) Iodine.
d) Astatine.

38) Kyoto protocol is related to

a) Greenhouse gas production control.
b) Oil pollution control.
c) Control of CFCs production and use.
d) Control of Non-Biodegradable produce production.

39) Accumulation of non-biodegradable pesticides in the food chain in increasing amount at each higher tropic level is known as

a) Pollution.
b) Magnification.
c) Accumulation.
d) Biomagnifications.

40) The role and position of a species in an ecosystem is called

a) Ecological factor.
b) Ecological niche.
c) Ecosystem constant.
d) Biomass.

41) Which one does not come under the 3R method of waste management

a) Refuse.
b) Reuse.
c) Reduce.
d) Recycle.

42) Which of the following limits the number of trophic levels in a food chain

a) 10% energy is transfer to higher tropic level.
b) Diffusion of the food supply.
c) Water.
d) Unidirectional flow of energy.

43) Which level is mostly affected by bio magnification?

a) Producer level.
b) Herbivorous level.
c) Carnivorous level.

d) Apex level animals in the ecosystem.

44) World ozone day celebrated in

a) September 1.
b) September 16.
c) June 5.
d) September 5.

45) We should avoid use to disposable plastic bags due to

a) Use of toxic material during production.
b) Light weight and not reliable.
c) Made of non virgin plastics.
d) Made of non-biodegradable materials.

46) Ozone gas is

a) Mono atomic.
b) Diatomic.
c) Tri-atomic.
d) Tetra atomic.

47) Which will not have a self-sustained ecosystem?

a) Crop field ecosystem.
b) Pond ecosystem.
c) River ecosystem.
d) Forest eco system.

48) Identify the natural ecosystem

a) Pond.
b) Forest.
c) Lake.
d) All of the above.

49) The community of plants and animals that have common characteristics for the environment they exist in is called

a) Ecological niche.
b) Biomass.
c) Biome.
d) Bio magnification.

50) Eutrophication occurs due to the high presence of

a) Oxygen.
b) Nitrogen.
c) Chlorine.
d) Protein.

51) The decomposers in an ecosystem

a) Convert inorganic materials to simple form.
b) Convert organic materials to inorganic form.
c) Convert inorganic materials o organic form.
d) Do not breakdown organic compounds.

52) Aerosol pollutants mostly produced by

a) Aero plane.
b) Trains.
c) Automobiles.
d) All of the above..

53) All pants in the ecosystem are called

a) Phytoplankton.
b) Zooplankton.
c) Planktons.
d) Decomposers.

[A COMPLETE GUIDE TO MCQ]

CHAPTER:15

[Answers & Explanations]

1) **(c).** producers stored energy as chemical energy in food.

2) **(b).** Nektons are those animals in an ecosystem who could swim and navigates. Fish, Whales are the examples.

3) **(a).** Functional unit environment is ecosystem.

4) **(c).** biomass is the total quantity or mass in a given area. Abiotic are some bacteria and lower class animals and their presence in mass are very less compared to biotic components in an ecosystem such as plant, human, tiger & many more.

5) **(a).**

6) **(a).** If we want to protect apex consumer then we have to take care of trophic level of very next lower level and so on. By means of doing this we could originally protect the whole ecosystem.

7) **(a).** Lake and ponds are having a natural ecosystem.

8) **(c).** The food chain is always unidirectional as lower trophic level always eaten by higher trophic level but the reverse is not possible naturally. Snake always eaten by peacock nut peacock never eaten by a snake.

9) **(a).** Manure produces from composting of bio-waste materials.

10) **(c).** Small fishes are primary carnivorous as they eaten small insects of water.

11) **(c).** In a food chain the third level is always occupied by a carnivorous as the first level is always producers and the second level is herbivorous. Next trophic levels always take herbivorous as food so the third level should have carnivorous.

12) **(c).** the ozone layer is rebuilt by UV rays.

13) **(c).** Incineration is a waste disposal method. Mainly organic substances are combust by this method.

14) **(a).**

15) **(a).** ozone gas is the cause of all the diseases given in the question except Diarrhea.

16) **(b).** Cotton cloth, paper mold and stool are a natural waste and they are bio-degradable. Washing soda is non bio-degradable compound.

17) **(c).** Food web defines more than one source of food for a given trophic level from the different food chain. Peacock could not always get snake as a food but they could survive by eating grains, other reptiles etc. from the different food chain.

18) **(a).** 1%.

19) **(a).** Food web is multidirectional. Food chain and flow of energy have unique direction.

20) **(d).** The flow of material in an ecosystem is cyclic as all materials of nature is recycled back by biogeochemical cycle.

21) **(b).** food chains are basically two types.1. grazing food chain 2.detrius food chain.

22) **(d).** The **Montreal Protocol was on the Substances that Deplete the Ozone Layer** (a protocol to the Vienna Convention for the Protection of the Ozone Layer) is an international treaty designed to protect the ozone layer by phasing out the production of numerous substances that are responsible for ozone depletion in 1987.Kyoto (1992) protocol for control of greenhouse gases. Vienna protocol also for CFCs but it was in 1995.Copenhagen

protocol was released in 2009 on global warming.

23) **(a).** Refrigerant industries use CFCs which is the main cause behind ozone layer depletion. All other options are the creator of greenhouse gases.

24) **(a).**

25) **(d).** Producer stored energy =50000kJ. Grass hopper will get 5000KJ. Frog will get 500KJ and Snake will get 50KJ. By eating Snake Eagle will get 5KJ as per 10% rule of Lindeman.

26) **(a).** The parasitic food pyramid is inverted, because a single plant may support the growth of many herbivores and each herbivore provide nutrition to several parasites, which support many hyper parasites.

27) **(d).**

28) **(c).** An ecosystem includes all living and non-livings. Livings are different organisms, animals etc. Non-living things are the atmosphere, climate etc.

29) **(b). 90%** of the energy used by the preset trophic level organisms for their life process.

30) **(c).** In option 2 plankton will be eaten by fish and grass hopper both but human will not eat grass hopper but fish. So this will not constitute a food chain. In option 3 wolf, snake and tiger all are carnivorous so they will not eat grass and will not constitute food chain.

31) **(d).** food chain will demolish due to missing trophic level between the food chain. Here the number of grass hopper will increase rapidly and that will wipe out producers. Due to lack of food Snake and Eagle both will extinct slowly.

32) **(a).** CFCs are produced chlorine ions which will react Ozone gas and convert it to Oxygen and ClO. Then ClO again converted to Cl ion and O. Cl ion again attack Ozone and convert to oxygen.

$Cl + O_3 = ClO + O_2$

$ClO = Cl + O$

$Cl + O_3 = ClO + O_2$.

33) **(b).** Lindeman first proposed the 10% energy flow rule in an ecosystem. As per this rule in an ecosystem 10% of energy could transfer from one tropic level to next upper tropic level and it will never reverse back naturally. This energy efficiency maintains the finite trophic level in an ecosystem. Generally 4-5 trophic level exists in an ecosystem.

34) **(c).** CFC means Chloro Fluoro carbons

35) **(c).** Refer Q-35.

36) **(a)** Energy flow in an ecosystem is always unidirectional. Energy could never back to Sun again.

37) **(b)** Chlorine elements ion is the main cause behind Ozone layer depletion. Please refer Q-32.

38) **(a)** Please refer Q-22.

39) **(d)** Bio magnification is the accumulation of non-biodegradable pesticides in the food chain in increasing amount at each higher trophic level. Pesticides drain from agricultural land due to rain will accumulate in fish. Those fish will be eaten by a human being and in this way pesticide intake per unit food volume will be more for human beings.

40) **(b)** The role and position of a species in an ecosystem is called ecological niche.

41) **(a)** Refuse is the latest added concept to control waste generation.

42) (a) Only 10% of energy will flow from one trophic level to another trophic level control number of trophic level in a food chain. Maximum 4-5 number of the tropic level could exist in an ecosystem.

43) (d) Apex level consumer mostly affected due to bio magnification due to the concentration of pesticides.

44) (b). September 16 is celebrated as World Ozone day. On Sep16,1887 Montreal protocol was developed to protect Ozone layer depletion.

45) (d) Plastic bag completely degraded between 20 to 100 years.

46) (c) Ozone gas is Triatomic. O3.

47) (a) Crop field ecosystem is manmade ecosystem and unstable.

48) (d) Here lake means natural lakes only.

49) (c) Biome : The community of plants and animals that have common characteristics for the environment they exist in is called Biome.

50) (b) Eutrophication: the Excessive richness of nutrients (specially nitrogen based) in a water body due to run-off from the land. These cause the dense growth of plant life.

51) (b)

52) (a)

53) (a)

Management Of Natural Resources.

1) Total percent of the land area under forest cover in India is

a) 16
b) 19
c) 24
d) 28

2) Total percent of world land under forest cover is

a) 24
b) 31
c) 36
d) 41

3) Maximum number of an individual that can be suppored by a given environment is called

a) Carrying capacity.
b) Biotic potential.
c) Population size.
d) Environmental resistance power.

4) The demand for natural resources is _____ at an _____ rate with increase in population and improving technology.

a) Decreasing, log.
b) Increasing, exponential.
c) Constant, constant.
d) Increasing, linear.

5) Biogas generation is mainly based on the principle of

a) Fermentation.
b) Degradation.
c) Purification.
d) Incarnation.

6) The resources which are found everywhere are known as

a) Ubiquitous.
b) Non-renewable resources.
c) Renewable resources.
d) Man made resources.

7) Man made resources are

a) Renewable only.
b) Non-renewable only.
c) No resources are called man made resource.
d) Ubiquitous.

8) Ganga action plant came into action in

a) 1984
b) 1985
c) 1995
d) 1965

9) The Ganga action plan came in action to

a) Solve the issue of its distribution.
b) To improve the flow rate of ganges river.
c) To improve the quality of its water.
d) All of the above.

10) Choose incorrect statement(s) regarding advantages of ground water storage

1) Loss due to evaporation is nil.
2) Control mosquito breeding.
3) Provide moisture for vegetation.
4) Contamination during an earthquake is nil.

a) 1 & 2 are incorrect.
b) 2 & 3 are incorrect.
c) 3 & 4 are incorrect.
d) Only 4 is incorrect.

11) Coli form is

a) A bacteria.
b) A group of bacteria.
c) A group of algae.
d) Fungi.

12) The importance of small check dams are

1) Hold water for irrigation.
2) Hold water and prevent soil erosion.
3) Recharge ground water.
4) Hold water permanently.

a) All of the above.
b) 1 and 4 only.
c) 2 and 3 only.
d) 3 only.

13) A desirable level of coli form in Ganga as per Ganga action plant is

a) 10 MPN/100ml.
b) 100 MPN/100ml.
c) 500MPN/1000ml.
d) 500MPN/100ml.

14) The desirable limit of coli form in Ganges water is measured in MPN/ml. here MPN indicates

a) Minimum probable number.
b) Most probable number.
c) Minimum production number.
d) Most preferable number.

15) In present time 3R principle is converted to 4R principle in the waste management system. The new R is an indicator of

a) Reduce.
b) Refuse.
c) Recycle.
d) Reuse.

16) Which statement(s) is / are correctly explained the sustainable development of natural resources.

1) The growth of natural resources with minimum damage to the environment.
2) Hold the concept" environment first , development last".
3) Growth should acceptable to local peoples only.

a) 1 only.
b) 2 & 3 only.
c) 1 & 3 only.
d) All are correct.

17) Water pollution can be identified easily by

a) pH measurement.
b) BOD measurement.
c) Both (a) and (b).
d) Coli form measurement only.

18) Stake holders of forest product are categories in

a) 2 category.
b) 3 category.
c) 4 category.
d) 5 category.

19) Biodiversity hotspots mean

a) Where the number of species found is more.

b) Where the atmosphere is hotter than other area and having more number of species.
c) Where hot springs are mostly available.
d) Where a number of species available are less due to the hot atmospheric condition.

20) **Conservationist Mr. A. K Banerjee is related to**

a) Teak tree.
b) Sal tree.
c) Rose wood.
d) Mahogany tree.

21) **The method of growing more and more trees is called**

a) Agriculture.
b) Silviculture.
c) Apiculture.
d) Forestry.

22) **Bidis are prepared from**

a) Tendu leaves.
b) Bay leaves.
c) Basel leaves.
d) None of the above.

23) **Chipko andolon saves**

a) Sal tree.
b) Teak tree.
c) Eucalyptus tree.
d) No specific tree was considered for this andolon.

24) **Chipko Andolon was related to**

a) The killing of wild animals.
b) Felling of domestic trees.
c) Felling of forest trees abruptly.
d) Burning of forest abruptly for getting potash.

25) **Ground water will deplete due to**

a) Use of organic farming.
b) Earthquake.
c) Afforestation.

d) Hydro power plant.

26) **Cycling time of water is approximately**

a) 1 month.
b) 9 days.
c) 100 days.
d) 1 year.

27) **Which of the following is not the benefit of water harvesting**

a) Self-sufficiency to water.
b) Conserve ground water.
c) Reduce cost of pumping.
d) Reduce Hydro power per unit cost of generation due to more availability of water.

28) **Amrita Devi Bisnoi award is instituted for**

a) Conservation of specific endangered plant.
b) Conservation of Wildlife.
c) Conservation of any one natural resource.
d) Conservation of classical dance.

29) **Select the non eco friendly activity**

a) Using aerosol perfumes.
b) Using cycle for transport.
c) Using jute bags for shopping.
d) Using natural dyes for coloring cloths.

30) **The concept of "Biosphere Reserve" developed by**

a) UNDP.
b) UN.
c) UNESCO
d) Govt.Of India.

31) **Due to rising heights of dam following problems are created**

1) Large terrestrial phytoplankton and zooplankton submerged.
2) Large numbers of peoples require to relocate.

3) Power generation cost increase.
4) A large amount of agricultural land gets submerged which creates unemployment.

a) 1, 2 & 4 are correct.
b) 2, 3 & 4 are correct.
c) 1, 2 & 3 are correct.
d) All are correct.

32) Kulhs is related to

a) Water preservation by natural channels.
b) Water preservation by manmade channels.
c) Soil preservation by recycle soil made utensils.
d) Soil reservation from chemical fertilizers.

33) Most of the time it was observed that conflicts appeared between local peoples and government during construction of large Dams because

a) Social problems.
b) Economic problems.
c) Environmental problem.
d) All of the above.

34) "Switching off" fan when you are not in the room is one step towards

a) Refuse.
b) Reduce.
c) Reuse.
d) Recycle.

35) Tawa irrigation project is in

a) Maharashtra.
b) Madhya pradesh.
c) Bihar.
d) Orissa

36) Arabari forest of West Bengal is famous for

a) Teak wood.
b) Rose wood.
c) Sal wood.
d) Mahogany wood.

37) Mining is a big source of pollution because

a) It degraded the inner core strength.
b) It discards a large amount of slug for every ton of metal extraction.
c) It produces Carbon dioxide.
d) All of the above.

38) Reusing is the more efficient management of resources compare to recycle because

a) Recycle contaminated he produces.
b) Recycle use some energy to make it reuse able.
c) Recycle is not cost effective in most cases.
d) All of the above.

39) Which canal in Rajasthan bring greenery

a) Jawhar neheru canal.
b) Indra Gandhi canal.
c) Rajeev Gandhi canal.
d) Sardar sarovar.

40) By switching off the light we are ultimately saving

a) Money.
b) Natural resources.
c) Electrical equipment.
d) Tax.

41) During British era vast tracts of forest are cleared and a single species of plant (e.g. eucalyptus) was cultivated. This practice promotes

a) Biodiversity in the area.
b) The growth of the natural forest.
c) Economic growth of forest area.
d) Monoculture in the forest area.

42) In India a primary source of water is

a) River water.

b) Rain water.
c) Lakes.
d) Wells

43) Coli form is the cause behind

a) Gastric ulcer.
b) Gastroenteritis.
c) Pharyngitis.
d) Goiter.

44) Most rapidly degrading natural resource is

a) Forest.
b) Water.
c) Wind.
d) Petroleum.

45) All are the stakeholders of forest product except one. Find the different one

a) Local people.
b) Paper industry.
c) Wildlife conservative activist.
d) Iron industries.

46) Death of the last individual of a species is known as

a) Extinction.
b) Clad.
c) Diversity.
d) Evolution.

47) Which statement is wrong

a) Forest is a natural resource.
b) Forest conserves water.
c) There was no act for forest during British era in India.
d) Forest have great wild life diversity.

48) Amirata devi Bishnoi sacrificed her life for the protection of

a) Sal tree.
b) Teak tree.
c) Khejri tree.
d) Cactus tree.

49) Best fresh water pH level for plants and animals are

a) 2.5-5.5
b) 6.5-7.5
c) 7.5-9.5
d) 4.5-5.5

50) The success of forest conservation policy basically depends on

a) Protection of highest trophic level animals.
b) Protection of herbivorous and carnivorous both.
c) Protection of all physical and biological factors.
d) All of the above.

51) Social, economical and ecological equity is a necessary condition for achieving

a) Ecological development.
b) Economical development.
c) Sustainable development.
d) Social development.

52) Khadins, Ahars are ancient structures for

a) Conservation of soil.
b) Water harvesting.
c) Hydro power.
d) Grain storage before establishing cold storages.

53) Sanctuaries are made to

a) Rear animals.
b) Trap rare animals.
c) Protects animal.
d) Breed animals artificially.

54) Extinction of any species is recorded in

a) Blue data book.
b) Green data book.
c) Red data book.
d) Last data book.

55) Khadins water harvesting method is used in

a) Bihar.
b) Rajasthan.
c) Himachal Pradesh.
d) Maharashtra.

56) The cause behind abundant coli form bacteria in Ganges water is

a) Disposal of unburnt corpses into the water.
b) Discharge of effluents from electroplating industries.
c) Washing of cloth.
d) Immersion of idols.

57) Which of the following is not a natural resource.

a) Wind.
b) Gorilla.
c) Sal tree.
d) Museum.

58) Biosphere reserve conserve and preserves

a) Wild land and wild animals.
b) Endangered animals and plants.
c) Natural vegetation.
d) Both 1 and 3.

59) Panda award is related to

a) Conservation of forest.
b) Conservation of Wildlife.
c) Conservation of energy.
d) Conservation of water.

60) The best soil for plant growth is

a) Loamy soil.
b) Alluvial.
c) Clay.
d) Gravel.

61) Red Data Book is related to

a) Endangered species.
b) Non Endangered species.
c) Resources which completely exhausted by human beings.
d) Endangered plant species.

62) Electronic waste is the adverse effect of

a) Industry.
b) Agriculture.
c) Mining.
d) Transportation.

63) Which statement(s) is/are correct regarding bio diversity

1) Biodiversity refers to the different species of flora and fauna.
2) Biodiversity is less in the forest compared to the desert.
3) Biodiversity is less in forest compare to Ocean.
4) It refers to a total number of individual rather than the different number of different species.

a) 1, 2 and 3 are correct.
b) 1,3 and 4 are correct.
c) 1,2 and 4 are correct.
d) 1 and 3 are correct.

64) Which of the following is not the cause of modern agriculture system?

a) Eutrophication.
b) Bio magnifications.
c) Ozone depletion.
d) Ground Water pollution.

65) Supportive and assimilative capacity are the components of

a) Carrying capacity.
b) Holding capacity.
c) Capturing capacity.
d) Containing capacity.

66) Sustainable development concept was conceived in

a) 1970
b) 1987
c) 1990
d) 2000

67) Select incorrect statement regarding the Sustainable development

a) It is linked to environmental conservation including economic development.
b) Is not consider the view point of stakeholders.
c) Is a long and persistent development.
d) None of the above is incorrect.

[A COMPLETE GUIDE TO MCQ]

CHAPTER: 16

[Answers & Explanations]

1) **(c).** As per 2015 total % of the forest area of India was 22% but in 2018 it up by about 2%. Present % of forest cover is 23.8%.

2) **(b).** According to UN report 2010 World total forest cover area is 31%.

3) **(a).** A maximum number of an individual that can be supported by a given environment is called carrying capacity of that ecosystem.

4) **(b).** Natural resources demand increases exponentially with increase in population and improving technology.

5) **(a).** Biogas is produced from bio waste which bio degrades by means of anaerobic bacteria's i.e. fermentation process. This process is generally expedited at 38 to deg. C.

6) **(c).** As per 2015 total % of the forest area of India was 22% but in 2018 it up by about 2%. Present % of forest cover is 23.8%.

7) **(b).** According to UN report 2010 World total forest cover area is 31%.

8) **(a).** A maximum number of an individual that can be supported by a given environment is called carrying capacity of that ecosystem.

9) **(b).** Natural resources demand increases exponentially with increase in population and improving technology.

10) **(a).** Biogas is produced from bio waste which bio degrades by
means of anaerobic bacteria's i.e. fermentation process. This process is generally expedited at 38 to deg. C.

11) **(c).** As per 2015 report total % of the forest area of India was 22% but in 2018 it up by about 2%. Present % of forest cover is 23.8%.

12) **(b).** According to UN report 2010 World total forest cover area is 31%.

13) **(a).** A maximum number of an individual that can be supported by a given environment is called carrying capacity of that ecosystem.

14) **(b).** Natural resources demand increases exponentially with increase in population and improving technology.

15) **(a).** Biogas is produced from bio waste which bio degrades by means of anaerobic bacteria's i.e. fermentation process. This process is generally expedited at 38 to deg. C.

16) **(a).** The natural resources which are found everywhere are known as ubiquitous.

17) **(d).** Manmade resources are also called capital resources, are material riches which can be used for creating more wealth. Manmade resources are labor, small starting industries etc. These resources are ubiquitous.

18) **(b). Ganga Action plan** was formerly launched by then prime minister of India Mr. Rajeev Gandhi on 14 january,1986. As per CBSE syllabus book it has mentioned as 1985. It may be due to early expenses already carried out before launch GAP publicly. In this project center fund 100%.

19) **(c).** main purpose of GAP was to improve the quality of its water. Option a and b has no relation with GAP.

20) **(d).** First 3 statements are correct and ground water only contaminated during an earthquake as at that time its oscillate up and down randomly.

Management of natural resources MCQ (class X)

[A COMPLETE GUIDE TO MCQ]   CHAPTER:16

21) (c). Coliform bacteria are defined as rod-shaped Gram-negative non-spore forming and motile or non-motile bacteria which can ferment lactose with the production of acid and gas when incubated at 35–37°C.

22) (c). Check dams reduce the flow velocity by holding the water and so it prevents erosion of soil. Due to holding water in a place ground water also get recharged at that place.

23) (d). 500MPN as per GAP 1986.

24) (b). MPN means the **most probable number** of microorganisms per volume in a given sample. It is a statistical testing method.

25) (b). Refuse added in recent time to educate people to refuse the use of non-biodegradable materials specially plastics. Refuse will reduce the use of the non-biodegradable material in the environment.

26) (a). The growth of natural resources with minimum damages to the environment is called sustainable development. Its motto is development should not affect our nature badly.

27) (c). pH and Biological Oxygen Demand [The amount of dissolved oxygen needed (i.e. demanded) by aerobic biological organisms to break down organic material present in a given water sample at certain temperature over a specific time period] is the main measure of pollution level identification for water.

28) (c). The stakeholders are 1. People living in and around the forest. 2. Forest departments. 3. Small or big forest dependent industries 4. Wild life.

29) (a). it is a bio-geographic area that is rich in biodiversity. About 25% of biodiversity area of the world have lost till date.

30) (b). He was related to conserving Sal tree in *Arabari* forest range in west Bengal.

31) (a). Power generation cost will decrease if the height of the dam increase due to increase power plant efficiency.

32) (b). Kulhs are manmade canal irrigation system used in Himachal Pradesh long time ago. Somewhere still it exists.

33) (d). Social problem: Displace a large amount of pheasant and tribal without adequate compensation. Economic: used a huge amount of money without much employment generation. Environmental: Submerge large area and produce greenhouse gas methane. Destroy huge local biological diversity.

34) (b). Switching off fan reduce the loss of electricity which could be useful afterwards.

35) (b). tawa reservoir is located in Hoshangabad, Madhya Pradesh. It was completed in 1978.

36) (c). See Q-30.

37) (b).

38) (b) Recycle use some energy to make it reusable. So direct reusing of the product is efficient method compared to recycle.

39) (b).

40) (b). Switching off light reduce the use of natural resources which could be used further when the requirement arises.

41) (d). Cultivating the same plant after felling local domestic plants is called Monoculture forest culture.

42) (b). India is a monsoon depended nation.

[A COMPLETE GUIDE TO MCQ]

CHAPTER: 16

43) **(b).** gastroenteritis is the inflammation of stomach and intestine due to coliform bacteria.

44) **(b).** Water is the most rapidly degrading natural resource.

45) **(d).** The iron industry does not use forest product so it's not a stakeholder of the forest.

46) **(a)** Extinction is the death of the last individual of a species.

47) **(c)** First forest act was passed in 1865 during the British era.

48) **(c)** Amrita Devi Bishnoi sacrificed her life for the khejri tree in kejrali village near Jodhpur in Rajasthan in 1731.

49) **(b)** true fresh water is either slightly acidic(pH6.5) of slightly basic(pH7.5)

50) **(d)**

51) **(c)**

52) **(b)** All are an ancient structure for water harvesting. Khadins are famous in ancient Rajasthan and Aharas are famous in ancient Bihar.

53) **(c)** Sanctuaries are made to provide safe space for wildlife animals to protect them.

54) **(c) Red Data Book** is a list of endangered species maintained by IUCN since 1964.

55) **(b)**

56) **(b)** Disposal of unburnt corpses into Ganga is the main cause behind the excess formation of Coliform bacteria.

57) **(d)** Museum.

58) **(d)** Biosphere reserve concept was developed to conserve and preserve wild land, wild animals and natural vegetation.

59) **(c)** Panda award is Oscar equivalent award for the conservation of wildlife. Organized by WWF (World wide fund for nature; erstwhile known as World wild life fund).

60) **(a)** Loamy sand is the best soil for plants as its pores are food for ventilating air.

61) **(a)** See Q-54.

62) **(a)** Poisoning of soil is due to using chemical fertilizers during agriculture. Mining cleats slug as waste and transportation crease pollutants for air.

63) **(d)** Forest biodiversity is more compare to desert and biodiversity refers to any number of individuals present in the ecosystem for any number of species.

64) **(c)** Modern agriculture system uses chemical fertilizer and heavy water for sowing high yield seeds. Those chemical fertilizers got drained and create eutrophication, bio-magnification and also reduce the level of ground water etc. Ozone layer depletion occurs due to CFCs.

65) **(a)** See Q-3.

66) **(b).** 1980 It was first conceived from Brundtland Commission report.

67) **(b)** Sustainable development always takes care of stakeholders and conservation of natural resources with optimum uses of it for the betterment of mankind is the main motto behind sustainable development.

www.ingramcontent.com/pod-product-compliance
Lightning Source LLC
Chambersburg PA
CBHW062353220526
45472CB00008B/1785